*The First
Pictorial History
of the American
Oil and Gas
Industry 1859-1983*

Ruth Sheldon Knowles

The First Pictorial History of the American Oil and Gas Industry 1859-1983

Ohio University Press
Athens, Ohio
London

Books by Ruth Sheldon Knowles

The Greatest Gamblers: The Epic of American Oil Exploration
Indonesia Today: The Nation That Helps Itself
America's Oil Famine: How It Happened and When It Will End
America's Energy Famine: Its Cause and Cure

© Copyright 1983 by Ruth Sheldon Knowles
Printed in the United States of America.
All rights reserved.

Library of Congress Cataloging in Publication Data

Knowles, Ruth Sheldon, 1915-
    The first pictorial history of the American oil and gas industry, 1859-1983.

    Includes index.
    1. Petroleum industry and trade—United States — History. 2. Gas industry—United States—History.
I. Title.
HD9565.K599   1983   338.2′728′0973   82-22485
ISBN 0-8214-0693-0

Dedication

To the great American wildcatters, scientists and technologists who have dramatically proved, since the world's first commercial oil well in Pennsylvania in 1859, that there is an abundance of oil and gas in the earth, and that its development in the United States will continue for at least another century, as long as they are free to explore its huge potential and have the economic incentive to do so.

Contents

Acknowledgments	ix
In the Beginning	1
The Well that Changed the World: Birth of the Modern Oil Industry, 1901-17	19
Minds of Men: The Art of Petroleum Geology is Born	41
Winning World War I: "The Allies floated to victory on a sea of oil."	45
The Roaring Twenties	51
The Industry Comes of Age: 1930s	63
Winning World War II: "A matter of oil, bullets, and beans."	75
Energy Revolution: 1945-1970	89
Energy Crisis: 1970s	111
New Oil and Gas Horizons: 1980s and Beyond	139
Energy Future	163
Selected Bibliography	165
Index	167
About the Author	171

Acknowledgments

The wonderful thing about compiling and writing a pictorial history of the scope of this one, is the extraordinary cooperation I have received in assembling the most important, and rare, pictures to illustrate the industry's great achievements. I am exceedingly grateful to all the oil companies, service companies, libraries, universities and individuals who so generously made available their picture resources, which are acknowledged in the text.

In addition, I wish to particularly thank certain people who enthusiastically brought to my attention pictures I needed, but didn't know I needed, until they produced them. Richard Drew, of the American Petroleum Institute has organized the finest collection of historical and contemporary pictures of the oil and gas industry in the nation and his guidance and help was invaluable. I also have special thanks to give to Stafford Acher and Anita McGurn of Texaco, Bob Finney of Phillips Petroleum Co., John Steiger of Cities Service Oil Co., Manny Jimenez of Arco, and Princetta Johnson of Exxon. My love and thanks go to my daughter, and assistant, Carroll, whose help was indispensable. I am especially grateful for the dedicated and imaginative help of Patricia Elisar, Director of Ohio University Press, in making pictures and text flow together.

My finest source of inspiration, of course, has been the great oil and gas wildcatters, scientists and technologists, who have made America the world's greatest industrialized nation and whose imagination, courage and ingenuity have made the story of oil and gas the greatest romance in industrial history.

New York City   Ruth Sheldon Knowles

IN THE BEGINNING

..............the ancient earth bled oil, asphalt, and gas from its pores in seeps, springs, lakes, and pits. . . . Man from his earliest appearance on all the continents, marveled at the strange substances, called them by many names—slime, bitumen, pitch, asphaltum, tar—and found uses for them . . . light, mortar, waterproofing, medicines, weapons, and religious rites. . . .

- Following God's instructions on how to build an ark of gopher wood, Noah did "pitch it within and without with pitch"
- The Tower of Babel, and the walls of Jericho and Babylon were made of bricks "and slime they had for mortar"
- "And the vale of Siddim was full of slime-pits and the kings of Sodom and Gomorrah fled and fell therein"
- In order to save him, the mother of Moses "took for him an ark of bulrushes, and daubed it with slime and with pitch and put the child therein" and laid it in the river's brink where he was found by the daughter of Pharoah
- The "eternal fires" of Baku on the Caspian Sea—gas seeps ignited accidentally by lightning—inspired ancient religious fire-worshipping cults
- Egyptian, Babylonian, Syrian, Assyrian, Persian, Greek, and Roman rulers fought for many centuries for control of the Dead Sea asphalt seeps . . . Mark Anthony captured them and gave them to Cleopatra
- In A.D. 600 Confucius wrote about wells drilled in China for salt which found oil and gas
- In the Americas thousands of oil, gas, and asphalt seeps and springs were used by the Indians from time beyond memory for making medicine, caulking boats, waterproofing blankets, cementing beads, and creating decorative designs
- In A.D. 1188 Geraldis Cambrensis, a writer and naturalist, wrote about an oil seep in his *Itinerary through Wales:*

Not far hence is a rocky eminence, impending over the Severn, called by the English Gouldcliffe, or golden rock, because from the reflections of the sun's rays it assumes a bright golden color. Nor can I be easily persuaded that nature hath given such splendor to the rocks in vain, and that this flower should be without fruit, if anyone would take the pains to penetrate deeply into the bowels of the earth; if anyone, I say, would extract honey from the rock and oil from the stone. Indeed many riches of nature lie concealed through inattention, which the diligence of posterity will bring to light; for, as necessity first taught the ancients to discover the conveniences of life, so industry, and a greater acuteness of intellect, have laid open many things to the moderns.

It would be almost seven hundred years before Cambrensis's visionary belief would become reality. Man and his industries relied on wind, water, and wood for fuel until the seventeenth century when Europeans began to mine and use coal. As late as the 1860s, wood supplied three-fourths of the total fuel supply of the United States. However, the nineteenth-century Industrial Revolution accelerated a greater demand for lubricants than vegetable oils and sperm oil from the ocean's dwindling number of whales could supply. Also, whale oil for illumination was scarce, and lamp oil from coal was costly.

The time had come to penetrate the "bowels of the earth" to see what splendor nature had given the rocks. Despite all the signposts nature had left for mankind for thousands of years throughout the world, the world's economic history and civilization's rapid progress would be changed dramatically by a small, shallow hole bored in the rocks in backwoods Pennsylvania in 1859.

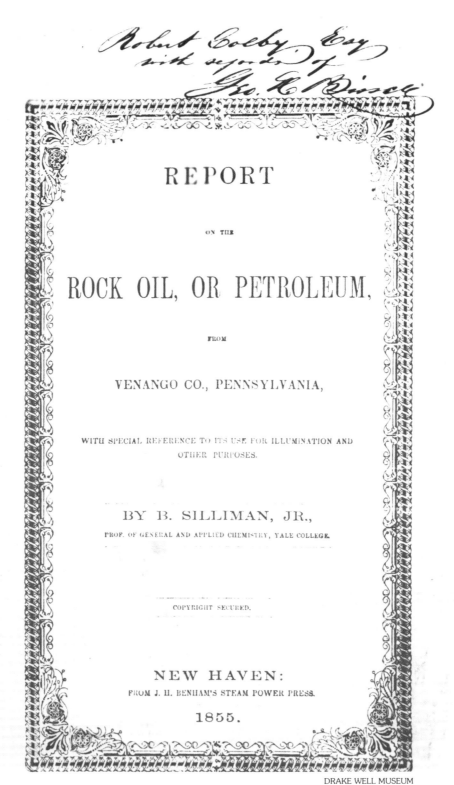

The above 1855 report, confirming that "rock oil" could be refined into kerosene to light lamps, inaugurated the world's first commercial search for petroleum.

AMERICAN PETROLEUM INSTITUTE

. . . . there was no money to continue pounding a hole into the rocks looking for oil near a spring on Oil Creek, Pennsylvania. . . . Colonel Edwin L. Drake, in top hat and frock coat, above, with Peter Wilson, a Titusville druggist who had endorsed a $500 bank loan for Drake to keep drilling, was gloomily certain he had failed. . . .

The unique idea to drill for oil had seemed to Edwin Drake to be a sound, sure way to make a fortune. An ex-railroad conductor, he had been hired by George Bissell, a New York lawyer, to direct the operations of the Pennsylvania Rock Oil Company. Bissell dubbed him Colonel to impress the local people.

Bissell, who had been given a sample of oil skimmed from the nearby spring, wondered if it could be used for lamp fuel in place of the standard illuminant, whale oil, which was becoming scarce. Benjamin Silliman, Jr., a Yale University chemistry and geology professor, analyzed the sample and reported it could be refined into excellent kerosene.

3

PETROLEUM, OR ROCK OIL.

A NATURAL REMEDY!

PROCURED FROM A WELL IN ALLEGHENY COUNTY, PA.

Four hundred feet below the Earth's Surface!

PUT UP AND SOLD BY

SAMUEL M. KIER,

CANAL BASIN, SEVENTH STREET, PITTSBURGH, PA.

The healthful balm from Nature's secret spring,
The bloom of health, and life, to man will bring;
As from her depths the magic liquid flows,
To calm our sufferings, and assuage our woes.

CAUTION.—As many persons are now going about and vending an article of a spurious character, calling it Petroleum, or Rock Oil, we would caution the public against all preparations bearing that name not having the name of S. M. KIER written on the label of the bottle.

PETROLEUM.—It is necessary, upon the introduction of a new medicine to the notice of the public, that something should be said in relation to its powers in healing disease, and the manner in which it acts. Man's organization is a complicated one; and to understand the functions of each organ, requires the study of years. But to understand that certain remedies produce certain impressions upon these organs, may be learned by experience in a short time. It is by observation in watching the effects of various medicines, that we are enabled to increase the number of curative agents; and when we have discovered a new medicine and attested its merits, it is our duty to bring it before the public, so that the benefits to be derived from it may be more generally diffused, but have no right to hold back a remedy whose powers are calculated to remove pain and to alleviate human suffering and disease. THE PETROLEUM HAS BEEN FULLY TESTED! About one year ago, it was placed before the public as A REMEDY OF WONDERFUL EFFICACY. Every one not acquainted with its virtues, doubted its healing properties. The cry of humbug was raised against it. It had some friends;—those that were cured through its wonderful agency. These spoke out in its favor. The lame, through its instrumentality, were made to walk—the blind, to see. Those who had suffered for years under the torturing pains of RHEUMATISM, GOUT and NEURALGIA, were restored to health and usefulness. Several who were blind have been made to see, the evidence of which will be placed before you. If you still have doubts, go and ask those who have been cured! Some of them live in our midst, and can answer for themselves. In writing about a medicine, we are aware that we should write TRUTH—that we should make no statements that cannot be proved. We have the witnesses—crowds of them, who will testify in terms stronger than we can write them to the efficacy of this Remedy, who will testify that the PETROLEUM has done for them what no medicine ever could before—cases that were pronounced hopeless, and beyond the reach of remediate means—cases abandoned by Physicians of unquestioned celebrity, have been made to exclaim, "THIS IS THE MOST WONDERFUL REMEDY EVER DISCOVERED!" We will lay before you the certificates of some of the most remarkable cases; to give them all, would require more space than would be allowed by this circular. Since the introduction of the Petroleum, about one year ago, many Physicians have been convinced of its efficacy, and now recommend it in their practice; and we have no doubt that in another year it will stand at the head of the list of valuable Remedies. If the Physicians do not recommend it, the people will have it of themselves—for its transcendent power to heal, *will* and *must* become known and appreciated—when the voices of the cured speak out; when the cures themselves stand out in bold relief, and when he who for years has suffered with the tortures and pangs of an immedicable lesion, that has been shortening his days, and hastening him "to the narrow house appointed for all the living," when he speaks out in its praise, who will doubt it? The PETROLEUM IS A NATURAL REMEDY—it is put up as it flows from the bosom of the earth, without anything being added to or taken from it.

DRAKE WELL MUSEUM

"The lame were made to walk . . . the blind to see."

THE BANCROFT LIBRARY

Benjamin Silliman, Jr., the Yale University chemistry professor who was the scientific "father" of the petroleum industry.

GULF OIL

Uncle Billy Smith, driller on the Drake well. "He saw something dark on the water."

Bissell leased the land at Oil Creek and, with the backing of New Haven bankers, formed a company to try to develop oil in quantities.

Bissell chanced to see a patent medicine handbill showing a wooden tower housing machinery used to drill salt wells. Samuel M. Kier, a Pittsburgh salt works owner, was selling rock oil that came up with the salt brine as a medicinal cure-all. If oil could be found indirectly by drilling for salt, why not use the same method to find oil directly?

Drake hired Uncle Billy Smith, a sprightly old blacksmith knowledgeable about salt wells, to drill the well. The enclosed derrick and shed contained a big wheel with coiled cable, which, powered by steam, would roll up the cable suspended from a pulley high in the derrick, then release it so that the iron bit on the end of the cable would strike powerful blows, pounding a hole through the rocks.

There was so much water in the upper soil layers that the hole continued to cave in. Drake conceived the idea of driving an iron pipe into the ground until it rested on solid rock at thirty-two feet and then began drilling the hole inside the pipe. His ingenious idea would be adopted by all hole drillers after him.

On Saturday, August 27, 1859, the well was sixty-nine and one-half feet deep. There was no oil and no money. After spending $2,490, the New Ha-

5

Those who could not afford drilling rigs in the new oil boom kicked holes down by the ancient Chinese springpole method. A rope with drilling tool attached ran over a pulley suspended in a crude derrick and was fastened to a pole. Kicking bent the pole down. Its natural spring moved the tool up and down.

The Pennsylvania oil boom rivaled the California Gold Rush ten years earlier. Soon wells were gushing as much as 4,000 barrels a day. Below, the John Benninghoff farm on Oil Creek was a typical scene. It also exploded the theory that producing wells could not be found on hills. The forest of wooden derricks bled the Pennsylvania earth of such quantities of oil that by the end of 1861 it sold for ten cents a barrel.

SHELL   AMERICAN PETROLEUM INSTITUTE

ven bankers refused to send more. Drake had kept the well drilling for a few weeks on borrowed money. He was despondent. On Sunday morning Uncle Billy, puttering around the well, was dismayed to see that water had seeped up to within ten feet of the top. He saw something dark on the water and lowered a can into the hole. It came up dripping with oil.

When Drake rigged up a pump, the world's first commercial oil well began producing thirty-five barrels a day, selling for $20 a barrel. Such was the unspectacular beginning of an industry soon to be trademarked by its gushers. But it sparked a rush to punch holes along Pennsylvania creek banks.

John Davison Rockefeller, a prudent young Cleveland, Ohio, produce merchant, was intrigued with the fact that a gallon of kerosene sold for twice as much as a barrel of crude oil. Cleveland was rapidly becoming the nation's leading oil-refining center. The coal-oil industry had pioneered technology for refining lamp kerosene from coal a few years before the Drake discovery. Now that an abundance

John D. Rockefeller at thirty-three in 1872. "When it's raining porridge you'll find John's dish right side up," said his elder sister, Lucy.

Rockefeller's Standard Oil Company of Ohio's number one refinery in Cleveland, 1870.

Family happiness in the mellow glow of the new Kerosene Age.

N.Y. PUBLIC LIBRARY

of petroleum could provide cheaper kerosene, petroleum refineries were mushrooming.

In 1862, Rockefeller helped finance a refinery. Three years later he decided to devote full time to refining. By 1870 he had two large refineries and incorporated the Standard Oil Company of Ohio. The nation, emerging from the Civil War, was headed for boom times. Europe was clamoring for American kerosene. Rockefeller shrewdly foresaw that control of the industry would belong to whoever could obtain a throttlehold on refining and transportation. Thus, by 1880, owning 80 percent of America's refining capacity and 90 percent of its pipelines, Standard Oil emerged as the largest and richest manufacturing company in the world.

Rockefeller's methods of achieving monopoly —secret rebates, gobbling up small companies, bribery, collusion, deception—were those of the times. However, he practically invented efficiency and cut the pattern for twentieth-century big business.

He was contemptuous of both the public and small businessmen. After he formed the great Standard Oil Trust in 1882, he proudly declared: "The day of combination is here to stay. Individualism has gone never to return." He had no understanding of the kind of individualism that prompted small oil producers along the Clarion River in Pennsylvania to state: "Resolved, that in the consideration of the position taken by the 'large producers' in the matter of the combination, the 'small producers' of Clarion County do hereby resolve that we, the said 'small producers' don't care a damn for the combination or for the 'large producers' or anybody else and that's what's the matter with Hannah."

Following Standard's lead, one industry after another combined to form monopolistic trusts. The Sherman Anti-Trust Act of 1890 resulted in response to a wrathful public's demand to be protected from exploitation, but it would be twenty-one years before the act would be used effectively to bust the trust. In the meantime, a series of basic inventions and discoveries would mark the end of an era and the opening of a spectacular new one.

DRAKE WELL MUSEUM

By 1862 the oil region was producing 3 million barrels a year and the pressure of storing and transporting it to markets where it could be refined demanded innovations. As shown in this stereoptican view, the first storage tanks were large wooden vats.

AMERICAN PETROLEUM INSTITUTE

Early transportation of oil was in barrels, which soon became the standard measurement for oil production. The barrels were loaded on wagons or piled on flatboats and floated down rivers and streams.

SOCAL

The idea of substituting bulk for barrels and shipping it by railroad inspired Amos Densmore, a pioneer producer. He designed wooden tanks, each carrying forty-five to fifty barrels of oil, mounted on flatcars, and shipped them by rail to New York in 1865. Although he tried to keep his innovation a secret, the railroads quickly adopted it for use throughout the region.

DRAKE WELL MUSEUM

In 1865, when Samuel Van Syckle, an inventive Jerseyite, began constructing the first crude oil pipeline, running 5 miles from Pithole to the Oil Creek railroad, he was ridiculed as well as threatened by teamsters. He buried the welded iron pipe joints two feet underground, below plough depth. When the pipeline began operating successfully at a charge of one dollar per barrel, hundreds of teamsters who had been getting three dollars for hauling a barrel over the same route, began tearing it up. The sheriff had to post guards. W. H. Abbott and Henry Harley bought the line and built the great Allegheny Transportation Company with 500 miles of pipe to various oil centers. At left, one of its terminals is shown.

9

The national excitement created by the discovery and development of oil in Pennsylvania inspired a half dozen popular songs between 1859 and 1864.

In the industry's first decade, uses of petroleum by-products, after refining kerosene from crude, multiplied so that an enthusiastic 1869 writer exclaimed: "It is something to know that a cargo of petroleum may navigate a river, cross a lake or ocean, in a vessel propelled by steam it generated, acting upon an engine it lubricates, and directed by an engineer who may grease his hair, anoint his body, perfume his clothing, enrich his food, rub his bruises, freshen his liver, and waterproof his boots with the same article."

N. Y. PUBLIC LIBRARY

Early kerosene illuminant peddlers also carried petroleum by-products "guaranteed to cure most of the diseases known to man."

AMERICAN PETROLEUM INSTITUTE

Actress Lillian Russell's portrait on an 1886 ad for a one-dollar bottle product with "elements required for hair to feed upon."

AMERICAN PETROLEUM INSTITUTE

Petrolina, made "from the heart of petroleum," came in eight varieties from curing bronchial tubes to dressing hair.

Cleveland, O., June 15, 1883.
Dear Sir:

We desire to call your attention to our Mineral Seal 300° Fire Test Burning Oil, which has gained for itself so valuable a reputation both for its great illuminating powers -- giving a brilliant white flame-- and for its wonderful safety properties, being superior to all other burning oils in this respect, withstanding a heat of 300° Fah. before igniting, for which reason it is especially adapted for use in Railway Coaches and Passenger Steam Boats. It is now being extensively used for this purpose all over the United States and Canada.

Very Truly,
Standard Oil Company.

Refining began as a "tea kettle" operation, heating crude to vaporize lighter elements and condense them into various products, principally kerosene. The lightest products, gasoline and naphtha, initially were dumped as waste. Soon, however, naphthas were used as solvents for dissolving India rubber; waterproofing materials; and in manufacturing paints, varnishes, furniture polish, and glues.

Gasoline's first use was in 1861 as a local anesthetic by a renowned Boston surgeon, Dr. Henry Bigelow. He then asked a refiner to develop a product with even greater freezing effects. He called the result "rhigolene," after a Greek word meaning "extremely cold." For almost two decades "rhigolene" was the primary local anesthetic for doctors and dentists nationwide. An even lighter gasoline product, known today as liquified petroleum gas, was used in compression machines to manufacture ice.

AMERICAN PETROLEUM INSTITUTE

N. Y. PUBLIC LIBRARY

An 1883 kitchen marvel—the Golden Star multiburner oil range. Price—twenty dollars. Tin oven—four dollars extra.

Vaseline (from the German wasser, *or water, and Greek* alaion, *or oil) became popular worldwide. At first people feared a "petroleum jelly" might explode.*

N. Y. PUBLIC LIBRARY

California's first commercial well in Pico Canyon pumped twenty-five barrels a day from 370 feet in 1876. The longest-lived well in the West, it was still pumping a barrel a day in 1982.

SOCAL

The most important early gasoline use was for making air-gas to improve pipeline distribution of gas made from coal for illumination in cities and factories. Coal gas was too expensive for any but the upper classes. However a new "naphtha" gas from gasoline competed with it, and by 1873 had replaced coal gas in major cities.

Petroleum lubricants became the most valuable refined by-product when refiners learned how to deodorize them. They rapidly became an international best-seller.

Paraffin wax made from crude oil had the widest and most diversified use of any petroleum by-product. It was used primarily for candles. But the second largest national market was typically American—paraffin chewing gum, which was "highly recommended for constant use in ladies' sewing circles." Paraffin wax uses quickly spread—sealing jellies, preserving meat, adding an attractive sheen to bonbons and cake fillings, sealing beer barrels and wine casks, coating pills and splints on fractures, making impressions for false teeth, coating wrapping and writing papers, and preserving wood.

Another major breakthrough was Robert Chesebrough's 1869 discovery of how to make petroleum jelly which he trademarked "Vaseline." To the present day it is the most widely used petroleum by-product in pharmaceutics. Its unique characteristics made it a superior base for salves, unguents, pomades, lotions, and other pharmaceutical preparations. It was also valuable in currying, stuffing, and oiling all kinds of leather and as a lubricant for machines.

Below, Lyman Stewart, at age nineteen, a tanner's son, caught Drake well fever in 1859 and invested in a Pennsylvania well, then lost the $125 he had saved to become a missionary. He never recovered from the disease and in 1892, after countless dry holes, sparked California's first big oil boom.

UNION OIL

Mining prospectors, seeking oil dug a shaft 370 feet deep on a Los Angeles street corner and produced seven barrels a day in 1892. Above the Los Angeles City field by 1895 was the state's biggest producer with "wells as thick as the holes in a pepper box."

"California will be found to have more oil in its soil than all the whales in the Pacific Ocean," Professor Benjamin Silliman, Jr., reported enthusiastically in 1864. "The oil is struggling to the surface at every available point and is running down the rivers for miles and miles."

Thomas A. Scott, Lincoln's Assistant Secretary of War, and vice-president of the Pennsylvania Railroad, had sent Silliman, whose report on Pennsylvania's rock oil inspired the drilling of the Drake well, to investigate stories about southern California oil seeps. After drilling six wells and finding only a little heavy oil in one of them, Scott and fellow investors gave up. The search for oil languished for a decade.

In 1876 two former Pennsylvanians, Denton C. Scott and Robert C. McPherson, brought in California's first commercial well at a depth of 370 feet, producing 25 barrels a day, in San Fernando Valley's Pico Canyon. It did not start a boom, but it lured Lyman Stewart and W. L. Hardison, Pennsylvania oil producers, to come to California to try their luck. It was consistently bad. California's oil was

Newhall refinery, near Los Angeles, one of California's earliest, in 1895.

Carl Benz, a German mechanical engineer, invented the first gasoline-powered vehicle—a three wheeler—in 1885. In 1893, Charles E. and J. Frank Duryea created the first gasoline-powered American automobile. Charles had the idea and Frank built the machine—a one-cylinder gasoline engine with electrical ignition, installed in a secondhand carriage. Charles is shown in his gasoline "buggy" in 1895, with the first set of pneumatic tires developed by the Duryeas. That year it became famous when it won the first automobile race in America—54 miles—sponsored by the Chicago Times Herald.

SOHIO

hard to find. It was also heavy, thick and yielded little kerosene. Stewart and Hardison managed to produce 200 barrels a day by 1890 and formed the Union Oil Company. The whole state was only producing 1,000 barrels daily.

California's oil prospects seemed to be insignificant, but Lyman Stewart, the team's wildcatter, kept Union exploring. In 1892, in a canyon northwest of Los Angeles, one of Stewart's wildcats flowed 1,500 barrels of oil a day over the derrick, like a fountain, down the canyon into the Santa Clara River and out into the ocean. Lyman Stewart had fulfilled Professor Silliman's vision of a quarter of a century before.

California prospectors streamed to the hills with the old-time abandon and enthusiasm of Gold Rush days. But the next strike was bizarre. In Los Angeles, Edward L. Doheny, a mining prospector,

In 1896 Barnum and Bailey featured the Duryea's automobile as the star attraction of their circus.

AMERICAN PETROLEUM INSTITUTE

seeing an ice wagon loaded with tar, asked the driver about it. He learned the tar had been dug from nearby pits in an area named "Brea"—the Spanish word for tar. The ice plant used it for fuel, and Los Angeles adobe buildings were roofed with it. Doheny visualized a mother oil pool feeding the pits. He excitedly sent for his mining partner, Charles Canfield, and they spent their last dollar buying a corner lot. Never having seen an oil well, with pick and shovel they sank a shaft, 4 by 6 feet, and continued digging. At 165 feet their picks opened a layer of rock that flowed seven barrels a day.

The shallow oil "mine" caused even wilder excitement than Lyman Stewart's gusher of a few months before. There was a stampede to buy and lease lots. Canfield and Doheny led the herd. There were so many frantic diggers and drillers, a dismayed city council outlawed oil drilling in city limits. Each operator promptly announced he was drilling for water. Within two years, 3,000 "water wells" were producing oil, and the Los Angeles City field was the state's biggest producer.

Lyman Stewart, Edward Doheny, Charles Canfield—these pioneers literally had only scratched the surface. By 1903 their continued efforts would make California the nation's leading oil-producing state.

As the nineteenth century ended, more than one-half of the nation's 76 million people were living on farms with no more freedom of movement than a horse could give. America's 4,000 horseless carriages, powered by steam, electricity, or gasoline were still just playthings for the rich.

Although no one realized it, the Kerosene Age was ending. The Fuel Oil Age was beginning as more oil was being discovered outside Pennsylvania. In 1880 Pennsylvania produced 28 million barrels—99 percent of America's oil—and was the world's only source of an export surplus. But in 1900 Pennsylvania produced only 20 percent of America's 63 million barrels. Oil had been found in fourteen other states. However, illuminating oil was still the most important product.

The ideas of young men creating new machines would usher in the Gasoline Age. Their simultaneous discoveries and inventions would do more for the complete change and advancement of American industry during the first fifteen years of the twentieth century than had been done during the whole century before. Free to dream and to pursue these dreams, they would give Americans freedom of movement and create opportunities for wealth for the many instead of the few.

In order for this to happen as quickly as it did someone had to prove that there was an abundance of oil in the earth for all and for all purposes.

A young man in Texas was about to do that.

In 1904 when Standard Oil Company (Indiana) introduced its first tank wagons, few realized that the horse was an "endangered power species." Harness shops were still big business on America's main streets.

STANDARD OIL (INDIANA)

OIL & GAS JOURNAL

Backyard tinkerer, Henry Ford, shown above in his first automobile, in 1896, dreamed of quantity production, until on January 1, 1901, Detroit newspapers carried a dissolution notice of his Detroit Automobile Company with sale of all its assets. Ford was broke and discouraged. In Dayton, Ohio, two youthful bicycle repairmen, Wilbur and Orville Wright, were equally discouraged in their efforts to develop a flying machine. "Man won't be flying for a thousand years," Wilbur concluded in disgust.

Spindletop, the astounding southeast Texas gusher which changed the history of America and the world on January 10, 1901. Flowing 100,000 barrels a day from a depth of 1,020 feet, in a solid black stream 175 feet in the air, it proved that there was an abundance of oil in the earth.

THE WELL THAT CHANGED THE WORLD:
Birth Of The Modern Oil Industry, 1901-1917

Every curious thing in nature attracts to it a curious mind. Patillo Higgins, a Beaumont, Texas, jack-of-all-trades, visiting the eastern oil region in the 1880s, recalled Big Hill, a low circular mound near Beaumont. Sulfurous gas seeped from its crest. If oil and gas seeping from Pennsylvania springs led to finding oil, why not in Texas, too?

Higgins persuaded Beaumont businessmen in 1892 to finance three shallow wells on Big Hill, but the cable tools could only drill 400 feet. A proper hole could not be made through the mound's heaving quicksands. Meanwhile in 1894, a well drilled for water at Corsicana, 200 miles northeast, struck oil. In a few years it was the first commercial oil field in Texas, producing 1,000 barrels a day. But it was just a curtain raiser.

After seven frustrating years, Higgins could find no more backers to test Big Hill. He advertised his dream in a manufacturing journal. He received only one answer, but that was enough. It was from Captain Anthony F. Lucas, an Austrian mining engineer, who had been developing salt mines in Louisiana from salt domes mushrooming to the surface. He had found some oil and sulfur connected with salt and thought Higgins's mound might be a salt dome which could have oil associated with it. Lucas leased the land, giving Higgins a 10 percent interest.

In mid 1899 Lucas confidently started drilling with a light rotary rig he had used in Louisiana drilling for salt. Water well drillers developed the rotary principle in the 1870s as a faster way of going through soft formations, and it was used extensively nationwide. A rotary, or revolving platform, gripped a pipe with a bit fastened to its end. As the pipe rotated, the bit ground a hole down instead of pounding it down as cable tools did. Corsicana saw the

AMERICAN PETROLEUM INSTITUTE

Captain Anthony F. Lucas, Austrian mining engineer. He answered a magazine advertisement to drill at Spindletop and became the father of the modern oil industry.

first successful use of rotary drilling for oil. Before steam engines were used, power came from a mule harnessed to the rotary platform and circling the well.

Lucas was only able to drill 575 feet through the quicksand with his light rotary equipment when gas pressure collapsed the pipe. However, he had recovered a few gallons of heavy green crude oil from the hole. He had no more capital to get better equipment to drill another well. Seeking financing, he approached Standard Oil Company. Its renowned production expert, Calvin Payne, visited Big Hill and, despite the oil show, reported that there was "no indication whatever to warrant the expectation of an oil field on the prairies of southeastern Texas."

A University of Texas geology professor, Dr. William Battle Phillips, came to the rescue. He told

19

Colonel James M. Guffey, left and John H. Galey, a famous, successful Pittsburgh oil-prospecting team, raised $300,000 for Captain Lucas to drill the Spindletop discovery well after he ran out of money on his first attempt.

Lucas he thought his theory was right and gave him a letter of introduction to John H. Galey, of Pittsburgh. Galey and his partner, Colonel James M. Guffey, were nationally known successful oil prospectors. Galey was the oil finder of the team, and Guffey was the politician and money raiser. On Galey's recommendation they agreed to invest $300,000 if Lucas would lease all the acreage on the mound and around it. He leased about 15,000 acres, receiving an eighth interest. He proudly refused to accept any salary which soon meant that in order to eat while the well was drilling, Mrs. Lucas would cheerfully sell their furniture and use egg crates and apple boxes for chairs and tables.

Colonel Guffey immediately borrowed $300,000 from his friend, Andrew W. Mellon, the Pittsburgh banker. Big Hill was in business again. Galey hired the Hamill brothers, Al and Curt, leading Corsicana drillers, to drill the well under Lucas's supervision. The Hamills tackled Big Hill with a heavier rotary rig, but the difficulties encountered were so challenging that by the time the well was finished the Hamills and Lucas had discovered and perfected the most important techniques of modern drilling for oil.

Trying to master the quicksands, the Hamills discovered that if, instead of pumping water down the hollow pipe to keep the bit cool and force the cuttings up the outside of the pipe, they pumped mud it plastered the sides of the hole, making them

Andrew W. Mellon, Pittsburgh banker and later, Secretary of the Treasury, who loaned Colonel Guffey the $300,000 to invest in the Lucas well, little knowing that it would be the basis for his building the Gulf Oil Corporation, his greatest money maker, exceeding his huge businesses of banking, aluminum and carborundum and coke processing.

The first successful use of rotary drilling for oil was the first commerical Texas field, Corsicana, in 1894, using mule power to grind a hole through the rocks.

SMITHSONIAN

firm. To make thicker mud they drove a herd of cattle through their water pit. Thus began today's multimillion dollar specialized drilling mud industry.

At 880 feet they hit solid rock and had a show of oil. On Galey's recommendation they decided to continue drilling. They could always come back to the oil show and try to produce it. At 1,020 feet the bit hit a crevice. The pipe turned helplessly. The Hamills changed bits and had lowered about 700 feet of drill pipe down the hole when the well began to spout mud. They watched in terror as four tons of heavy pipe shot out of the hole and over the derrick, its joints breaking as they went skyward. They ventured back to the rig floor to start shoveling mud when, with no warning, there was a deafening roar and the well erupted like a volcano—mud, gas, oil, and rocks shooting hundreds of feet in the air. Quickly dousing the boiler fire, they sent a helper to bring Captain Lucas from town.

The roar had shaken the countryside. The black geyser could be seen for miles. The town of Beaumont emptied, with everyone heading for Big Hill on horseback, in buggies, and by foot.

Patillo Higgins was the last to know. He had left town that morning. His name was absent on the towering granite shaft erected forty years later, commemorating the site of "the first great oil well in history—the Anthony F. Lucas gusher." But Higgins unwittingly gave the oil field its popular name. He told reporters he was going to drill a well on Spindletop Heights, some real estate acreage he had bought on the hill. The name caught their fancy. There were other Big Hills. The giant discovery needed a different name. Spindletop was just the thing.

As the great well gushed 100,000 barrels a day in a solid black stream 175 feet into the air, Texas railroads advertised: "In Beaumont, You'll See a

EXXON

This picture was long thought to be the rotary rig that drilled the Lucas well, but no contemporary photographs show a building in the background. Obviously, the arrow meant it was the driller's first well in the Spindletop field.

This was the daily scene at the Beaumont railroad station during the insane Spindletop boom. Excursion trains brought 15,000 people a day to see the wonder. Thousands of fortune hunters spent millions of dollars for leases. There was such a need for cash that a principal shipment into Beaumont was freight cars loaded with silver dollars. The boom became known as "Swindletop." More money was lost than made.

Gusher Gushing." Within two days 10,000 sightseers came to marvel. The Hamill brothers managed to cap the well in nine days. At first the well was thought to be a freak, but within three months others drilled five more gushers, each equaling the Lucas well. The news flashed around the world that six wells in Texas could produce as much in one day as all the oil wells in the whole world.

Lucas had leased only from large landowners. Countless small owners, each with a few acres, provided the opportunity for the most insane speculators' boom the world had ever seen. Forty thousand fortune seekers, swindlers, gamblers, and prostitutes descended on Beaumont's 10,000 residents. Millions of dollars changed hands in hours. Oil was so plentiful that it sold for three cents a barrel; drinking water was so scarce it sold for five cents a cup. When doctors announced it was safer to drink whiskey than Beaumont water, the local Women's Christian Temperance Union indignantly dispensed free boiled water.

By year's end there were 440 gushers on the hill. Oil was pouring out as though through a sieve.

Nobody knew the field was being badly damaged. Millions of barrels would be irrecoverable as the gas pressure needed to bring the oil to the surface was being released all at once. When Lucas visited the field three years later, he was saddened by the devastation. Of the 1,000 wells drilled, only 100 were producing so much as 10,000 barrels a day. "The cow was milked too hard," he said, "and, moreover, she was not milked intelligently."

At the celebration of Spindletop's fiftieth anniversary, Dr. E. DeGolyer, another great oil finder, paid the hill its truest tribute. "I once traveled over the countryside of Cape Briton," he said. "It was a hard land. 'What does this country produce?' I asked. The reply was from a dour Scot. He looked me squarely in the eye and replied, 'Men.' And so it was with Spindletop. It was a producer of men."

Men from Spindletop forged a dynamic new art of oil prospecting. Inspired oil explorers were ready to go anywhere and everywhere now that Spindletop had shown there was an abundance of oil in the earth.

Spindletop also gave birth to some of the na-

GULF REFINING COMPANY
PORT ARTHUR, TEXAS

GULF OIL

tion's greatest oil companies—Gulf, Texaco, and Sun. No longer did Standard Oil Company have a monopoly on oil, although indirectly it would benefit as much as the others by acquiring properties discovered by independent oil finders whose roots were in Spindletop

When Spindletop erupted, Colonel Guffey urged the Mellons to form a big oil company. It took two more gushers to convince them to finance the Guffey Petroleum Company and the Gulf Refining Company. Guffey bought out his partners, Galey and Lucas. When Spindletop's production dwindled, the Mellons bought Guffey's interests. Their new Gulf Oil Corporation was well on its way with a refinery at Port Arthur, southeast of Beaumont on the Gulf of Mexico; tankers; markets at home and abroad; and oil exploration in other states.

Joseph S. Cullinan, a Pennsylvanian-turned-Texan, was the principal developer of the Corsicana field and owned a small refinery there. He came to Spindletop at the beginning of the boom, organized the Texas Company, backed by eastern capital, to buy oil to sell to eastern refineries. The company's cable address was Texaco. The Texaco star was born in January 1903, when it drilled a tremendous gusher on land it held at Sour Lake, 20 miles from Spindletop. Within two years this salt dome field was producing more than Spindletop and has been so prolific that new producing horizons at depth are still being discovered in the 1980s.

Below, left, Joseph S. Cullinan, founder of the Texas Company. Its first big gusher in 1903 was next to the Sour Lake Hotel, whose lobby is shown below right. Guests came to be cured of practically anything by bathing in its sulfurous waters bubbling with gas.

TEXACO INC.

AMERICAN PETROLEUM INSTITUTE

Horses and carts travel down the plank road called Boiler Avenue, above, which provided access to Spindletop's boiler locations. Wells were drilled so close together that a man could walk from one rig floor to another without ever touching the ground. Every well was a gusher.

When speculators came to Spindletop, if a gusher wasn't being drilled in, they never left disappointed. Stock promoters of the many companies on the hill would open up wells and let them gush.

EXXON

TEXACO INC.

Fire raged for a week in Spindletop's first great disaster, September 1902, shown above. A well ignited from a cigar carelessly discarded by a driller. The well was gushing above the derrick top when the flame reached it. There was no chance to close it in as valves had not been installed.

This lake of oil held 3 million barrels of oil and was the common and dangerous way of storing oil in Spindletop's early days.

FRED A. SCHELL

Shoestring Alley at Sour Lake, right, was like Boiler Avenue at Spindletop. No one realized what irreparable damage was being done to the new fields by such close drilling which permitted gas needed to produce oil to escape all at once.

TEXACO INC.

Spindletop, proving that vast quantities of oil and gas could be found around salt domes, sparked the rush to find and drill around other salt domes. By 1903 Gulf Coast salt domes were producing 20 percent of the nation's oil.

Most importantly, the flood of oil had swept away old industrial methods. Railroads, ships, and factories, using this cheaper, more efficient fuel, were multiplying as never before. The Fuel Oil Age had begun.

Below, the Texas Company's first refinery at Port Arthur, 1903, together with Gulf's refinery, established Port Arthur as a major refining and transportation center as ocean-going ships and tankers could dock there.

TEXACO INC.

Spindletop and Sour Lake crude oil and refined products for the industrialized eastern seaboard and Europe initially traveled from Port Arthur in majestic sailing ships such as The Arrow *shown here in 1903, right. However the ships were soon replaced by iron tankers, using fuel oil for power, such as the J. M. Guffey, below, built in 1902, first of Gulf's fleet.*

OIL & GAS JOURNAL

GULF OIL

Above, the streets of Port Arthur were paved with asphalt in 1908. Asphalt was Texaco's first marketed product.

Rotary drilling was revolutionized in 1909 when Howard Hughes, Sr., a Texas oil prospector, invented the cone roller bit. It drilled a straight round hole quickly and efficiently through hard rock. Below, original plant of the Hughes Tool Company. The company created the fortune inherited by Hughes's son, Howard, Jr., aviation and movie pioneer and enormously rich, eccentric recluse.

ETHYL CORPORATION

The new abundant oil era captured the nation's fancy, inspiring stories in national pulp weeklies, telling of virtue's triumph in the oil fields. Other pulps were True Blue, Starry Flag Weekly, Do and Dare Weekly, *and* Brave and Bold.

Adventure weeklies flourished nationwide. The Beaumont oil strike was immortalized in Secret Service.

There was a great pipeline race, 1905 to 1907, to get Oklahoma's kerosene and gasoline rich crude in the huge Glenn Pool discovery. One pipeline tied into Standard's lines to the eastern seaboard in the longest pipeline network in the world. Oil flowed in Gulf's and Texaco's lines from Oklahoma to Port Arthur, Texas. Left, early pipeliners were called "tong gangs" because of heavy tongs used to handle and screw pipe joints.

AMERICAN PETROLEUM INSTITUTE

Oil is found in abundance only if a great many people are looking for it at once in all sorts of unlikely places. The nation's oil finders were proving this.

Between 1900 and 1904, Texas and California doubled the country's oil production. In 1903 California became the leading oil producer.

In Oklahoma, still Indian territory, a gusher near Tulsa opened the rich Glenn Pool in 1905. The new boom brought a rush of prospectors. By 1907, at statehood, Oklahoma was the nation's biggest oil-producing state.

Glenn Pool started a pipeline race. Unlike the heavy fuel crudes of Spindletop and Kansas, Glenn Pool oil was rich in kerosene. Standard Oil immediately began laying a pipeline from Kansas and was the first to reach Oklahoma. In 1906 Glenn Pool oil flowed to the Atlantic coast through the world's longest pipeline network.

Spindletop's wells had practically ceased to flow. Gulf Oil Company's Port Arthur refinery and fleets of tankers were almost idle. Gulf started building a 550-mile pipeline from Port Arthur to Oklahoma. Gulf was also enthusiastic about the gasoline potential of Glenn Pool's light crude to sell to France where the automobile industry was already flourishing. The company felt the same development was about to happen in America.

Texaco also wanted to expand refining and producing operations and began building its own Port Arthur-Glenn Pool pipeline. The Gulf and Texaco pipelines finished in a dead heat in 1907.

Of all the rugged new breed of wildcatters, the most outstanding was Harry Ford Sinclair, of Independence, Kansas. He had been a pharmacist in his father's drugstore until the discovery of oil near the town showed him his career. He was the epitome of the great oil gambler. His first producing well in Kansas made him determined to outdo John D. Rockefeller. Glenn Pool gave him his big chance. When he heard of the new strike he went there so quickly that he was able to buy some of the choicest leases before prices rocketed. Oil pools, big and small, were found in an ever-widening circle from Glenn Pool. Drilling was cheap, for oil was found at shallow depths. Sinclair would drill a choice lease, prove it for oil, then sell it, often to Standard, to get money for bigger plays. Harry Sinclair was building an oil empire.

ATLANTIC RICHFIELD

Harry Ford Sinclair began his meteoric career as a great wildcatter in Oklahoma's 1905 Glenn Pool Boom. Above, shown with John Overfield in an electric car, his discoveries would help usher in the Gasoline Age.

By 1904 everyone was automobile mad. There were 178 automobile factories employing 12,000 workers. "Everyone can afford a Fordmobile," Henry Ford proclaimed. In 1903, eleven years after making his first car, he started again at forty with his third company and by 1904 had sold 658 Model A's at $750 apiece. The car was "built for business or for pleasure," his advertisements stated. "Built also for the good of your health—to carry you 'jarlessly' over any kind of half decent roads, to refresh your brain with luxury of much 'outdoorness' and fill your lungs with the 'tonic of tonics'—the right kind of atmosphere." In 1908 Ford announced he would henceforth produce his Model T—a car "built for the multitude." In 1909 in a population of 90 million, 200,000 Americans owned automobiles. In 1913, there were 1,260,000 cars registered.

AMERICAN PETROLEUM INSTITUTE

Outside assembly line at the Ford Highland Park, Michigan, plant in 1914.

SHELL

Above, the Automobile Gasoline Company of St. Louis in 1905 opened America's first drive-in service station with gravity tank and garden hose.

"Get a horse" was a familiar taunt to the ears of motorists stranded along the highways of the early 1900s, such as the one below. The first census of American roads in 1904 revealed that 97 percent of the country's 2 million miles of roads were dirt. What were called "improved" highways were mainly gravel. Soon, however, the highway system would expand and be paved with asphalt.

GENERAL MOTORS

PAUL H. GIDDENS

Above, a 1916 automobile station in North Dakota where the horse still had equal billing.

AMERICAN PETROLEUM INSTITUTE

Above, underground storage of gasoline brought about the curbstone pump in 1918. The driver filled the tank, operating the pump by hand.

STANDARD OIL (N.J.)

"Automobiling" created a market for new products made from oil. John Wanamaker's department stores in Philadelphia and New York sold automobiles in addition to all the appropriate clothing to wear in them.

AUTOMOBILING is a **severe tax** upon delicate complexions. Not alone the rush of air caused by the speed of the machine which has a tendency to drive the particles of dust from the road deep into the pores, but the exposure to varying conditions of weather, extreme heat, sunshine or bitter cold. These conditions may, however, be rendered entirely harmless to the skin of even the most delicate texture merely by the use of Daggett & Ramsdell's **"Perfect Cold Cream."** Before leaving the house rub a small quantity well into the pores of the face, neck and ears and remove what remains on the surface with a soft towel.

The dust and grime which adheres to the skin after your ride is finished, may then be removed by a second careful application. It will be found that not only has no injury resulted to the skin from exposure but a velvety softness and healthy glow will follow. To thoroughly enjoy the sport of automobiling, one's pleasure must not be spoiled by fears of a tanned or chapped skin. No lingering doubt nor dread need be felt if the above directions are carefully heeded.

Orville and Wilbur Wright made man's first powered flight in 1903 at Kitty Hawk, New Jersey, above. In 1909 New York saw its first airplane. Wilbur Wright rose from Governor's Island in his awkward flying box and circled the head of the Statue of Liberty while all the craft in New York harbor piercingly cried a salute to America's realization that man was finally as free in the air as he was on the ground. Two years later, C. P. Rogers made the first transcontinental flight from New York to Pasadena.

The military possibilities of flying machines were quick to be realized. Below, only eight years after the flight of Kitty Hawk, the U. S. Navy had its Curtiss Hydro, one of the earliest seaplanes. Lt. J. B. McClaskey of the U. S. Marines is the pilot.

In the race between California and Oklahoma to become the nation's leading oil producer, California jumped ahead in 1910. Lyman Stewart's Union Oil Company agreed to finish drilling a well for Lakeview Oil Company in return for 51 percent of its stock. Union's driller, Charles Lewis Woods, was known as "Dry Hole Charlie" because of never bringing in a producer. When he came to work on March 14, 1910, the earth erupted. Oil gushed in a huge stream over the derrick. Above the roar, "Dry Hole" Charlie was heard yelling, "My God, we've cut an artery down there!" Right, the gusher was greater than Spindletop with 125,000 barrels a day. It produced a record 9 million barrels, and "Dry Hole" Charlie became "Gusher" Charlie. Lyman Stewart had his greatest success, and California prospectors drilled with renewed excitement.

Below, in 1902 the Union Oil Company revolutionized ocean oil transportation with the *Whittier*—the first tanker with engines in the stern, reducing fire hazards, and cargo holds as an integral part of the hull. It was built in San Francisco for $195,000. All modern tankers have followed the design.

The charismatic Theodore Roosevelt became President in 1901 upon the assassination of William McKinley. The new century's first decade was an

UNION OIL   UNION OIL

EXXON

era of reform and "muckraking." Roosevelt initiated "trust-busting" of big business combinations. His principal crusade was against the great Standard Oil Trust and John D. Rockefeller.

In 1904 Standard Oil owned barely one-sixtieth of America's total oil production. However, it refined 84 percent of the country's illuminating oil, it held 86 percent of the export trade in illuminants, its 88,000 miles of pipeline transported 90 percent of the crude of older fields, and it controlled more than 88 percent of the sales of illuminating oil to American retailers.

Roosevelt's campaign was so effective that in 1909 federal courts ordered the Standard Oil Trust to be broken up into thirty-eight companies, a decision upheld by the Supreme Court in 1911. Rather than being "busted," Rockefeller became richer than ever through his stock ownership in all the companies, which prospered more individually than they had under his control. America was learning that competition and new ideas, rather than monopoly, were the ways to build a greater nation.

Although Rockefeller was one of the most vilified Americans in history, he was its most philanthropic. He channeled his great riches into many philanthropies which he organized as efficiently as he had his oil business. In 1913 he chartered the Rockefeller Foundation: "To promote the well-being of mankind throughout the world."

In 1913 a sleeping giant awakened in Oklahoma when the mammoth Cushing oil boom began. The field was discovered in 1912 by a Pennsylvanian, Tom Slick, but was considered a small one until Harry Sinclair drilled a 1,500-barrel gusher in 1913, and a few months later 5,000-barrel gushers were being found at 2,600 feet. Oil prospectors and workers streamed into Oklahoma by the thousands. The field covered thirty-two square miles and contained almost half a billion barrels of oil. Like Spindletop, it was the source of many important new companies. Oklahoma wildcatter, J. Paul Getty, parlayed his Cushing leases into founding Getty Oil Company and ultimately became one of

John D. Rockefeller in his later years.

CULVER SERVICE

Fires were a common disaster in the giant Cushing, Oklahoma, field which boomed in 1913.

This 55,000-barrel oil tank at Cushing was struck by lightning and burned, 1914.

the world's richest men. Between 1912 and 1919 Cushing was the source of 3 percent of the world's oil. The Cushing boom at its height was sheer insanity. Boom towns mushroomed the length of the field. Drumright was the hub with later towns dubbing themselves Dropright, Gasright, Alright, Downright, Justright, and Damright. Drilling was incessant as was crime and violence.

Cushing's most unusual visitor during its peak was Dr. Wilhem Wunsdorff, personal geologist of Kaiser Wilhelm. "I have never been so impressed with the possibilities of any oil country as I have been in Oklahoma," he marveled. Ironically, he was praising what would be a key to his country's defeat in a few years. When World War I began, one of the first bombs ever dropped by an airplane on a ship was a German bomb that struck a Standard Oil tanker. The ship was the *Cushing*.

Although oil was providing fuel power for the world, it couldn't have been found and developed without mule and horse power to haul equipment. This fourteen-mule team is crossing Tiger Creek Bridge on the road between Cushing and Drumright, the noisiest ten miles in America in 1913. It was never empty, day or night—a constant stream of teams and wagons hauling boilers, smokestacks, cable tools, still pipes, and lumber for derricks and stores. It was a pandemonium of cursing, shouting drivers; cracking whips; groaning wagons; clanking machinery. On hot days horses would drop dead under the strain.

This burning gas well at Dropright was a typical scene in early oil fields. With no market for natural gas, anyone unfortunate enough to find gas instead of oil simply let the well burn.

The grueling physical demands of working on a cable tool rig in an early boom field were compounded by the workers' capacity for whiskey. A typical Cushing cable tool oil worker, such as those shown above, scribbled his schedule on the back of an envelope:

UNIVERSITY OF TULSA, McFARLIN LIBRARY

11:00 A.M.	Get up
11:00–11:30	Sober up
11:30–noon	Eat
Noon to midnight	Work like hell
Midnight–3:00 A.M.	Get drunk
3:00–3:30	Beat hell out of them that's got it coming
3:30	Go to bed

MINDS OF MEN:
The Art of Petroleum Geology is Born

Cushing changed oil history as spectacularly as Spindletop had done, but not because of the amount of oil found. Oil men had been finding new pools for fifty-four years by following seeps, trends, hunches, divining rods, "doodlebuggers," and spirit mediums. In 1913, Charles N. Gould, founder of the University of Oklahoma geology department and the Oklahoma Geological Survey, presented a paper at an international geological congress stating that all Oklahoma's big pools, including Cushing, lay under anticlines, places where rock strata arched, dipping or inclining in opposite directions from a ridge, like the roof of a house. The following year the survey published a structure map of Cushing's surface geology, showing that the huge production was coming from a big structure whose arching could be seen and mapped.

A rush to find anticlines began. Oil companies established geological departments. Universities began offering four-year petroleum geology courses. Any doubts about the value of the new science were dispelled when Gould brought about the first major achievement of petroleum geology—the discovery of two giant oil fields in Kansas in 1914 and 1915.

Charles Gould, working for Cities Service Company, mapped an anticline near Augusta, Kansas, in 1913. Cities Service leased 10,000 acres along the anticline and found a 50-million-barrel oil field. This was the first field to be mapped geologically and leased before drilling began. Cities Service immediately expanded its geological staff, and within a year they had mapped a huge anticline near El Dorado, seventeen miles from Augusta. This mammoth covered thirty-four square miles, more territory than the great Cushing field. Cities Service leased 30,000 acres. The well discovered in 1915 opened up one of America's giant oil fields—275

"Any anticline is worth drilling," was the maxim of the "covered wagon geologist," Charles N. Gould, shown below, with beard, white shirt, and tie. His geological research launched the art of scientific prospecting.

OKLAHOMA UNIVERSITY LIBRARY

Cities Service geologists, 1915, who mapped the giant El Dorado, Kansas, anticline before drilling began—the first big oil strike where one company owned almost the whole field.

A nonboom town, Oil Hill, Kansas, in 1917, built by Cities Service for its workers to develop the geologically discovered El Dorado field.

million barrels. But there was no boom. Cities Service owned all but three and a half square miles of the field. Never before had one company owned almost all of a new discovery because never before had there been a way to judge the extent of a potential oil field in advance of drilling.

In early days, oil was thought to be in underground "pools." Geologists determined that oil and gas are organic matter formed in the earth and found mainly in marine rocks—successive layers of sand and mud deposited as sediments over the floors of former seas. Deeply buried layers harden into rock and the muds become impervious to fluids. But in the sandstones and limestones remain pore spaces which are occupied by fluids—oil and seawater—and gas. When they are trapped by impervious rocks they form oil "pools." When porous rocks are tapped, the oil is pushed to the surface by gas. Hence, the early gushers.

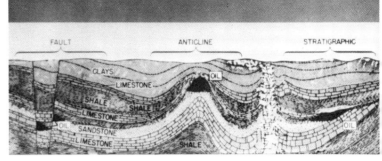

THREE PRINCIPAL TYPES OF OIL TRAPS
AMERICAN PETROLEUM INSTITUTE

Early petroleum geologists, nicknamed "rock hounds," learned to recognize surface clues indicating how buried rock strata formed various types of traps for oil and gas as shown above.

As Spindletop proved, salt plugs are potential oil traps beneath the cap rock and on their flanks as shown below.

AMERICAN PETROLEUM INSTITUTE

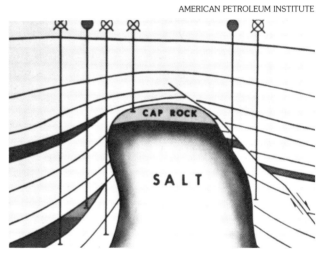

Below, a 1917 geologist in Model T equipped with surveyor's instruments.

TEXACO INC.

AMERICAN PETROLEUM INSTITUTE

Oil does not lie in pools but in the pore spaces of sandstone and limestone as shown at right.

43

The Texaco oil tanker S.S. Illinois sinks beneath the Atlantic after being torpedoed by a German submarine on March 18, 1917, less than one month before the U.S. declaration of war. The crew escaped in lifeboats and were rescued by British pilot boats. This photograph was taken from the German submarine and found on postcards being sold in Paris.

WINNING WORLD WAR I:
"The Allies floated to victory on a sea of oil"—
Lord Curzon, British Secretary of State for Foreign Affairs, 1918.

Even before its entry into World War I in 1917, America was supplying 90 percent of the great flow of oil which was the margin of victory. Storage tanks were drained and wells opened wide to ship 100,000 barrels a day to Europe. The new giant fields in Oklahoma and Kansas supplied most of it. Even so, there was not enough for both the war and domestic needs. "Gasless Sundays" were decreed.

The use of oil had revolutionized warfare as drastically as the invention of gunpowder. It was a war of fast ships, planes, tanks, trucks, and even taxicabs, a fleet of which had saved Paris in 1914 by rushing troops and supplies to repel invaders. Verdun was saved in 1916 by 40,000 motor trucks and tractors which transported troops and munitions to the fortresses.

Oil played a decisive role on the seas. The Allied supply line was almost severed in 1917 by German sinkings of Allied merchant marine ships. Winston Churchill called this "the gravest peril we faced in all the ups and downs of that war." The U. S. Navy's quick building of oil-burning submarine chasers, submarines, destroyers, and other vessels, with quick turnabout time, greater speed, maximum cargo space, and smaller crews, won the race against German sinkings.

The end of the war increased rather than diminished the need for new oil discoveries. In 1914 there were only 1,600,000 passenger cars in America. By 1918 there were 5,600,000. Farm tractors had increased from 17,000 in 1914 to 44,000 in 1918. Motor trucks had jumped from 85,000 in 1914 to 525,000 in 1918. The end of the war marked the real beginning of the aviation industry and the powering of all the world's ships by oil.

A few months before the end of the war in 1918,

The French Army going to war in taxicabs in 1914, below, to save Paris and meet the invading Germans on the Marne.

AMERICAN PETROLEUM INSTITUTE

ATLANTIC RICHFIELD

Major and independent oil company leaders worked together for the first time as the National Petroleum War Service Committee, 1917.

First row, seated, left to right: A. C. Bedford, *chairman of the committee, and chairman, board of directors, Standard Oil Co. of New Jersey;* M. L. Requa, *director, Oil Division, U. S. Fuel Administration;* J. W. Van Dyke, *president, The Atlantic Refining Co.;* George S. Davison, *president, Gulf Refining Co.;* T. A. O'Donnell, *director of production, Oil Division, U. S. Fuel Administration;* H. L. Doherty, *president, Cities Service Co.;* E. W. Clark, *president, Union Oil Co. of California;* H. E. Felton, *president, Union Tank Line;* and R. D. Benson, *president, Tide Water Oil Co.*

Second row, seated, left to right: E. C. Lufkin, *vice-chairman of the committee, and president, The Texas Co.;* Martin Carey, *general counsel, Standard Oil Co. of New York;* Edward Prizer, *president, Vacuum Oil Co.;* Ivy Lee, *director of publicity of the committee;* Samuel Messer, *president, Emlenton Refining Co.;* J. C. Donnell, *president, The Ohio Oil Co.;* A. G. Maguire, *director, jobbers and licenses, Oil Division, U. S. Fuel Administration;* W. P. Cowan, *former president, Standard Oil Co. (Indiana);* H. M. Blackmer, *president, Midwest Refining Co.;* J. S. Cosden, *president, Cosden & Co.*

Third row, standing, left to right: W. C. Teagle, *president, Standard Oil Co. of New Jersey;* J. E. O'Neil, *president, Prairie Oil & Gas Co.;* H. F. Sinclair, *president, Sinclair Oil & Refining Co.;* J. H. Markham, Jr., *Tulsa, Okla. producer;* George W. Crawford, *president, Ohio Fuel Supply Co.;* J. Howard Pew, *president, Sun Oil Co.;* C. C. Smith, *assistant secretary and treasurer of the committee;* J. A. Moffett, *secretary of the committee;* J. F. Guffey, *president, Natural Gas Association of America;* J. H. Barr, *president, National Supply Co.;* Frank Haskell, *president, Mid-Continent Oil & Gas Association;* W. S. Farish, *president, Gulf Coast Oil Association;* M. J. Byrne, *president, Independent Oil Men's Association;* J. A. Middleton, *Oil Division, U. S. Fuel Administration;* R. L. Welch, *secretary and general counsel, Western Petroleum Refiners Association;* and A. P. Coombe, *president, Standard Oil Co. of Ohio.*

SOUTHWEST COLLECTION TEXAS TECH.

The Merriman Baptist cemetery became nationally famous in the wild Ranger, Texas, boom in 1918–19 when Will Ferrell wrote a poem about the church's refusal to accept $100,000 for a lease.

> "All of oildom knows the answer
>   When the chairman shook his head
> Pointing past the men of millions
>   At the city of the dead.
>
> Why disturb the weary tenants
>   In yon narrow strip of sod?
> 'Tis not ours, but theirs the title,
>   Vested by the will of God.
>
> We, the board, have talked it over,
>   Pro and con, without avail,
> We reject your hundred thousand—
>   Merriman is not for sale."

With poetic license, Ferrell overlooked the fact that title to the cemetery was vested in its "weary tenants" not by the will of God, but by the will of a deceased church member, so nobody was available to legally sign a lease.

another spectacular oil boom erupted in north central Texas. Gushers in the adjoining Ranger and Desdemona oil fields seemed to be another huge bonanza. Major company and independent oilmen rushed in to spend millions. It was, perhaps, the most lawless of all oil booms as returning soldiers, brutalized by war, flocked to the fields for jobs. The pandemonium ceased almost as rapidly as it began. Within a year the gushers dried up. A 7,000-barrel-a-day gusher went dry in five months. This freak accumulation not only sobered the oil finders, but contributed to the growing belief that the nation had run out of oil. The war's drain on American oil fields had been too great, too fast. The nation was now consuming more oil than it produced. The threat of an oil famine was considered an impending national catastrophe. The nation panicked over the prospect.

47

ATLANTIC RICHFIELD

The cheerful young ladies below at Standard's Whiting refinery took over the jobs of the boys "over there" during the war. They packed and filled cans and helped the refinery produce, in 1918, 285 million wax candles, an oil by-product, to be shipped to France for the soldiers to use in the trenches.

STANDARD OIL (INDIANA)

PAUL H. GIDDENS

The greatest contribution of any oil company to the war was made by Standard Oil Company of Indiana. By 1909 the new automobile industry was consuming such an amazing amount of gasoline that it was in short supply. Refineries could only make a barrel of crude oil yield about 15 percent gasoline. Dr. William M. Burton, chemist, and manager of Indiana Standard's refinery in Whiting, Indiana, established the first experimental refining laboratory of any consequence in the industry and patented a process in 1913 which, by "cracking" the crude oil through higher pressures and temperatures, doubled its yield of gasoline. The first experimental tube still is shown above. "Cracked" gasoline from Indiana Standard refineries and its licensees increased cracked gasoline from 2 million barrels in 1914 to almost 12 million barrels in 1918, which averted a critical gasoline famine during the war years.

ATLANTIC RICHFIELD

The first of the 1920s' giant new oil fields was the Burbank, Oklahoma, field discovery well, above, in May 1920, made by Marland Oil Company on Osage Indian tribal lands. It was bigger than Glenn Pool and almost as big as Cushing. It catapulted many oil companies to enormous success, including Ernest W. Marland, one of the industry's most colorful wildcatters, and later, as Governor of Oklahoma, a pioneer of oil conservation.

THE ROARING TWENTIES

War production made America the world's greatest industrialized country. Incredibly, however, between 1920 and 1922 it became a net oil importer, drawing on new fields that American and British explorers had recently developed in Mexico. Congress quickly passed a depletion allowance tax incentive and opened the public domain for leasing. The government urged Americans to develop oil properties abroad. However, Britain, Holland, and France, also realizing oil's importance, were determined to keep Americans out of the Middle East where the British were developing Iran's oil, and out of the Far East where the Dutch were developing oil in the Dutch East Indies, now Indonesia. Iraq, which had been part of Turkey, was known to have great oil prospects. The League of Nations, to which the United States did not belong, gave oil rights to the Iraq Petroleum Company, jointly owned by the British, French, and Dutch. Only extreme American government pressure enabled a group of U.S. companies to win 23.75 percent interest in Iraq Petroleum Company. The Dutch buckled under in the Dutch East Indies when the U.S. government refused to let their American subsidiary lease Indian oil lands in Oklahoma.

When American companies began their foreign oil drive in 1919, the bulk of their $400 million investment abroad was in refining and marketing. By 1925, forty American companies invested $1 billion in exploration and development in Central and South America, the Far East, and Africa. However by then, fears of domestic shortages had vanished. Tax incentives and high oil prices triggered such intense, successful exploration at home that gushers became the trademark of the Roaring Twenties. By 1923 production had doubled from 1918. By 1929 it had tripled and passed the annual production mark of 1 billion barrels for the first time.

When the Osage Indian tribe was forced to sell its lands in Kansas in 1872 and buy a reservation elsewhere, their chief, Wah Ti An Kah, urged them to buy 1.5 million acres in the blackjack-covered hills of eastern Oklahoma instead of going to the western part of Kansas where there were still buffalo herds. "White man will not come to this land," he declared. "White man does not like country where

The famous "Million Dollar Elm," Pawhuska, Oklahoma, under which Osage Nation oil leases were auctioned off by Colonel E. E. Walters, in striped shirt below. Marland Oil Company bid as high as $1,990,000 for one 160-acre tract.

PHILLIPS PETROLEUM

PHILLIPS PETROLEUM

Frank Phillips, in headdress, cofounder of Phillips Petroleum Company, is made an honorary chief of the Osage Indian tribe, for his pioneer part in developing their oil-rich Oklahoma tribal lands.

Mexia, Texas, discovery well, 1920—the field discovered on the wrong geological idea that prompted the discovery of a great new geological tool in oil finding.

EXXON

there are hills. White man cannot put iron thing in ground here. This will be good place for my people." He was right in thinking this land would not attract the white man's plow. He hadn't counted on that other "iron thing," the drilling bit. When the Osages accepted allotment of their lands in 1906 to be ready for statehood, they reserved oil, gas, and mineral rights for the tribe as a whole to be shared proportionately. This eventually made them some of the richest people in the world.

Small pools were found on Osage lands in the early 1900s, one of which was discovered by the Phillips brothers, Frank and Lee, who founded the Phillips Petroleum Company. But the Osage became the nation's most important oil-producing area in 1920, when E. W. Marland discovered the huge Burbank field. Marland, a Pennsylvania lawyer, who considered himself an English gentleman, came to Oklahoma to seek an oil fortune. He was a flamboyant character and an early believer in the fledgling science of geology. His team of young geologists mapped the Burbank field and guided him to one successful Oklahoma field discovery after another. The great oil company he built—and ultimately lost—was the basis of today's Conoco.

Col. Albert E. Humphreys, an Oklahoma wildcatter, drilled a presumed anticline at Mexia, Texas, in November 1920, and found oil. No one knew how important the field was, but this was new oil territory in south central Texas. It would soon prove to be the basis of a dramatic breakthrough in the art of prospecting due to the deductive reasoning of a brilliant, young Kansas geologist—Wallace Everette Pratt.

Pratt had joined the Humble Oil & Refining Company in 1918 as its first geologist. This group of independent producers was an offspring of the Spindletop era. The company prospered in the short-lived Ranger boom when Standard Oil of New Jersey, shorn of its producing properties by the dissolution of the Standard Trust, bought a half-interest in the company. Immediately following the Mexia discovery Pratt spent a week in the field, establishing that the supposed anticline was actually a fault, a fracture in the rock strata caused by earth movement, forming an underground trap for oil. Pratt told Humble that the real producing area of the field would be west of where everybody else, who

EXXON

"Where oil really is, then, in the final analysis, is in our own heads"—Wallace E. Pratt, the flying philosopher-geologist, above, who raised the profession of petroleum geology to an eminence and a dignity that it would not have otherwise attained.

thought it was an anticline, believed the limits of production would be. Furthermore, if they followed the fault line they would find production further north. Humble leased every available acre. Pratt was right. A spectacular new field lay seven miles along the fault, and the company's hasty leasing campaign gained them 10 percent of it.

A year later when Pratt presented his novel fault theory to the American Association of Petroleum Geologists, of which he was a founder, many

Town lot drilling in Huntington Beach, California, 1920, at left, which enriched New England farmers.

were still skeptical. Within less than a month Pratt proved it beyond doubt. A small new discovery on a supposed anticline was made thirty miles north of Mexia at Powell. Recognizing that this was another fault accumulation, Pratt applied the same rule and Humble managed to acquire one-third of the field with its 12,000-barrel-a-day gushers.

Although both Powell and Mexia quickly declined, they produced 172 million barrels in five years. Humble was now firmly established as the largest producing company in Texas. More importantly, the art of geology had a remarkable new tool to be applied to finding oil elsewhere.

Wallace Pratt was a philosopher-scientist who had a profound influence on the thinking of thousands of earth-scientists. "It is the genius of a people that determines how much oil shall be reduced to possession," he said, "the presence of oil in the earth is not enough. Gold is where you find it, according to an old adage, but judging from the record of our experience, oil must be sought first of all in our own minds. Where oil really is then, in the final analysis, is in our own heads!"

Wallace Pratt believed that "oil in the earth is far more abundant and far more widely distributed than is generally realized." But he was also firmly convinced that "the prime requisite to success in oil finding is freedom to explore and only slightly less imperative is freedom to develop and produce the oil once it is found." The freedom to explore produced what he called "the most precious heritage our experience as oil finders has bequeathed to us—our multitude of itinerant wildcatters. The entire American oil industry is but the lengthened shadow of the independent oil man, whose form and substance are stamped indelibly over its whole structure."

Following World War I, California geologists began to suspect that some of the long, low hills in the Los Angeles area might cover deep-seated, oil-bearing anticlinal domes. Standard of California geologists studied a seaside resort area, Huntington Beach, which was not overly popular because of foul tasting water and the stench of "swamp gas" from nearby bogs. The reasons for this became obvious when the company drilled a 2,000-barrel-a-day gusher in 1920, starting the rich Los Angeles Basin boom. For a short time it made California Standard the largest producing company in the country. Lease men had to go to Maine, New Hampshire, and

Another spectacular town lot drilling boom in 1921 on Signal Hill, California, in back of Long Beach, shown at right.

Vermont in search of owners of a great part of the field. Many years before, an enterprising encyclopedia salesman had bought the land for almost nothing from a disappointed real estate promoter and increased his sales by offering New England farmers a free lot in sunny southern California with every set.

D. H. Thornburg, a Shell geologist who had grown up in Long Beach, remembered the tilted rock beds he had seen as a boy and mapped an anticline on Signal Hill, which rose sharply from the beach. He recommended a well. Shell's chief geologist, Dr. W. Van Holst Pelekaan, who was out of the country, hastened back to stop the drilling as Union had already drilled a dry hole on Signal Hill. He arrived in time to receive the news that on June 25, 1921, the Signal Hill well was flowing 600 barrels a day, opening up a field containing 926 million barrels.

The greatest political scandal in the nation's history erupted like a gusher in 1924. Known as the Teapot Dome Scandal, from the name of a rock outcrop resembling a teapot on a Naval Petroleum Reserve in Wyoming, it had its roots in the worldwide oil shortage scare after the war. In 1920, a giant oil field was discovered on the Salt Creek, Wyoming, anticline. Naval Petroleum Reserve No. 3 was on the anticline's south end. The Navy feared its oil was being drained by private operators and it was also concerned about drainage in two California reserves. Nervous about fuel supplies, it wanted to lease its reserves for development with the option to take royalty payments in fuel oil. The government committed administration of the reserves to the Interior Department. Secretary Albert Fall invited Harry Sinclair, the nation's largest independent, to develop Teapot Dome, and Edward Doheny, the early California developer, to handle the California reserves.

Two years after contracts were signed, a jealous neighbor informed the Senate that ex-Secretary Fall had made lavish improvements to his New Mexico ranch. The 1924 Senate investigation was sensational. It revealed that Doheny's son had delivered to Secretary Fall $100,000 in cash "in a little black bag" and that Harry Sinclair had loaned him $25,000 which was never repaid. Democrats proclaimed the scandal would overthrow the Republican adminis-

AMERICAN PETROLEUM INSTITUTE

Teapot Dome, above, the rock outcrop in the Salt Creek, Wyoming, oil field, which gave its name to the 1924 scandal that rocked the nation.

tration and adopted little aluminum teapots as campaign badges. Nevertheless, Calvin Coolidge was elected President. Civil and criminal suits dragged out for six years. Fall was convicted of criminal charges to defraud the government, sentenced to a year in jail, and given a $100,000 fine. Doheny was acquitted in a sympathetic jury trial. Sinclair served six months in jail on two contempt charges—one for refusing to answer a question in Senate committee hearings, and the other for charges of jury shadowing in the criminal trial in which he was acquitted. Despite the fact that he had built America's seventh largest oil company, Sinclair's golden dream of being greater than John D. Rockefeller was forever tarnished. Furthermore, it was a great tempest in a small teapot. Instead of producing an estimated 130 million barrels, the field fizzled out after producing 2 million barrels.

Seventy thousand square miles of desert land in West Texas was known as a petroleum graveyard. Every wildcat drilled was a dry hole except one 10-barrel well. Two adventurous young Texans, Frank Pickrell and Rupert Ricker, leased 431,360 acres near Big Lake from the University of Texas which owned 2 million acres in the area. They raised some money in New York to drill a well. Pickrell, at the request of the Catholic investors, climbed to the top of the derrick and sprinkled rose petals to christen the well Santa Rita—the Saint of the Impossible. At 3,038 feet, it flowed 100 barrels a day in 1923, but they could find no company to believe they had a real field unless they proved it with a second well.

Mike Benedum, a famous Pittsburgh wildcatter who had opened big fields in West Virginia, Il-

TEXAS MID-CONTINENT OIL & GAS ASS'N.

The 1923 well above christened "Santa Rita"—the Saint of the Impossible—which brought the petroleum graveyard of Texas to life and made the University of Texas the richest in the world.

SAM T. MALLISON

Mike Benedum, right, the great wildcatter who opened up the rich Permian Basin in 1924. Then by drilling a "good faith" well that his assistant, Levi Smith, left, had promised, he found one of the world's richest oil fields.

linois, and Louisiana, took the gamble, agreeing to invest a million dollars to drill eight wells. Two were teasers, pumping a few barrels daily. Six were dry holes. The great wildcatter decided to drill one more, 200 feet east of Santa Rita. The petroleum graveyard of Texas came to life when the ninth well roared in for 5,000 barrels a day, opening up the great Permian Basin. Within the next thirty years explorers would find almost 10 billion barrels of oil in its 70,000 square miles. There were more giant oil fields lying hidden under this wasteland than would be found in any other single area in the United States.

Two years later Benedum, in partnership with the Ohio Oil Company, drilled four dry holes thirty miles southwest of Big Lake. He was preparing to abandon the area when his assistant, Levi Smith, told him that among the ranches he had leased for the ventures was one owned by Ira G. Yates, who was convinced there was oil under his land. Smith had told Yates they would drill a well. It wasn't written in the lease, but Yates said Smith's word was good enough. Benedum told Smith not to worry. If Ohio wouldn't agree to participate in drilling another well he would do it personally. It never occurred to Ohio that Benedum would drill a well just to keep somebody else's word, so the company agreed. If they hadn't they would have missed out on one of the world's great oil fields containing 2 billion barrels. The good faith well gushed 70,000 barrels a day. Other Yates wells flowed 200,000 barrels a day.

The Texas Panhandle became another vast boom area in 1926 when a 10,000-barrel-a-day gusher discovered a prolific oil field forty miles northeast of Amarillo. It was called the Borger field as an enterprising Oklahoma real estate promoter, Ace Borger, established a town there, naming it after himself. Actually, the first discovery in the Panhandle was a gas field in 1918 north of Amarillo drilled on the recommendation of Charles N. Gould, the covered wagon geologist, who had mapped some anticlines there. He predicted to the prospectors that they were likely to find gas rather than oil. He did not realize what a prophet he was. Twenty years would pass before he was given full credit for locating the discovery well for the world's greatest gas field—one huge L-shaped structure extending from the Texas Panhandle across the Oklahoma Panhandle and into Kansas, covering 5 million productive acres. Along its northern flank in Texas lay four giant oil fields containing more than 600 million barrels.

The rowdy boom town at Borger, Texas, in 1926, below, which consisted of a two-mile-long main street. The Borger field initiated the prolific Texas Panhandle discoveries.

AMERICAN PETROLEUM INSTITUTE

As the great new flood of oil was released from the earth, the major companies expanded refineries and built thousands of miles of pipelines. Service stations sprang up at every crossroads and on every corner, for the flood of oil was matched only by the endless stream of cars rolling off assembly lines. The twenties in America were the greatest boom period any nation had ever known, and the symbol of the time was the automobile. In 1929, 4,587,000 new passenger cars were produced, a record that would not be surpassed for twenty years. There now were more than 23 million cars and 3,379,000 trucks and buses on the road, four times as many as there had been in 1919. As refineries poured out millions of gallons of gasoline to power them, they also met the phenomenal demands of ships, factories, tractors, home furnaces, and airplanes. A new aviation age was ushered in by the "Lone Eagle," Charles A. Lindbergh, and his spectacular solo 1927 transatlantic flight.

Everyone thought Oklahoma had seen its best days. There had been so much drilling that obviously no more giant oil fields were to be found. However, Cities Service management, needing to find new reserves, told its geologists to find a large untested area with good possibilities and low-cost

PHILLIPS PETROLEUM

The opening, above, of a Phillips Petroleum Company station in Wichita, Kansas, 1927, was the typical 1920s American scene.

These eager treasure hunters are not digging for oil as they did in Los Angeles streets in 1892, but are part of California Shell's famous "Treasure Hunts" in the mid twenties to find certificates good for merchandise prizes in buried plaster-of-paris shells. Competition to sell gasoline was fierce.

SHELL

The world hero, Charles A. Lindbergh, refuels his famous plane, "The Spirit of St. Louis," below, at Rockwell field, San Diego, California, prior to his 1927 transatlantic flight.

SOCAL

EXXON

Seminole in 1927 was as mad a drilling rush as the state ever witnessed. Everyone drilled wells at once—even drilling into each other underground. Mud was a special part of the madness. As shown above, wagons had to be used because automobiles and trucks mired down, as shown below, in the road leading out of Seminole.

EXXON

The fifteen-year period between Cushing and Seminole was more than a passage of time. It was a growing up period. Although early Seminole days were rough and violent, the old ways would no longer do. Cities Service, realizing the magnitude of its discoveries, built a model camp in the Seminole area, shown at right, with residences, bunkhouses, mess hall, recreation hall, doctor's office, ice plant, garage, children's playgrounds, and tennis courts. Other companies followed suit.

leases. The geologists settled on a blank spot of twenty square miles, fifty miles south of Cushing, which was part of the former Seminole Indian reservation. It was controversial geologically as it was highly faulted, but the geologists felt there might be oil at depth. At 3,751 feet the first well flowed 1,170 barrels a day. Another company drilled an 8,000-barrel-a-day gusher one mile southwest. Another giant had been found. Cities Service had leased conservatively. There was plenty of room for the 30,000 boomers who descended on the field. It was like a rerun of an old movie of the Cushing boom. In the twenty square miles blank spot there were five giant oil fields containing over 800 million barrels. Cities Service discovered three of them. Within a year, the Greater Seminole fields were producing 10 percent of the nation's oil.

Seminole discoveries were only a prelude. In scouring for prospects, Cities Service geologists found evidence of structure near Oklahoma City, the center of the state. Despite the many shallow dry holes drilled in the area, the company had such confidence in its geologists that it leased a 6,000 acre-block and in December 1928, its 6,335-foot wildcat blew in, running wild for almost 5,000 barrels a day from Oklahoma's most prolific producing sand, the Wilcox. The cannon that Oklahoma City officials fired when the well drilled in saluted the discovery of Oklahoma's greatest field and the nation's eighteenth most productive, containing 750 million barrels. As it developed, stretching out for eleven miles, it spread through the city's east side. It

CITIES SERVICE

59

The Oklahoma City field, above, discovered by geologists in 1928—the state's biggest field.

The Oklahoma City field became world famous in 1930 when a well being drilled on the Mary Sudik farm blew out, shown at right, spewing 20,000 barrels of oil and 200 million cubic feet of gas a day, blanketing the city with oil and gas. For eleven days, before it was brought under control, it was an international sensation with Floyd Gibbons, the famous war correspondent, broadcasting twice daily from the "Wild Mary Sudik" on nationwide radio hookup. Thousands of acres of oil-soaked land had to be ploughed under.

marked the end of the boom town era in Oklahoma. Even before the field reached the city limits, Cities Service hired a municipal planning engineer to work with the city government for an orderly development program.

As the 1920s ended, it seemed incredible that the decade had begun with fear of an oil shortage. Oil production had tripled. In 1929, more than 1 billion barrels flowed from the wells. Forty-two giant oil fields (100 million barrels or more) were found in Texas, California, Oklahoma, New Mexico, Arkansas, Wyoming, and Louisiana. In ten years more giants were found than in the previous twenty. As geologists looked at the record, they saw that three-fourths of the discoveries had been due to their efforts.

PHILLIPS PETROLEUM

After Lindbergh's flight America took to the air. Only 3,290,000 gallons of gasoline were consumed by U. S. civil aircraft in 1926. But in 1930, 28,531,000 gallons were used by the aviation industry. Above is one of the first aviation gasoline service trucks. Below, noted flyer Art Goebel skywrites for Phillips Petroleum Company in 1930.

PHILLIPS PETROLEUM

In East Texas, 8,000 people waited for two days in October 1930 to see if the well, shown above, would find oil. When it finally gushed, it had discovered the fabled 5-billion-barrel East Texas field.

THE INDUSTRY COMES OF AGE: 1930s

No wildcat well in America's history involved the faith, hopes, and dreams of so many thousands of people as did Dad Joiner's well in East Texas in 1930. And no well, until one in Alaska, thirty-eight years later, ever found so much oil.

Dad Joiner, at seventy, had been a promoter and driller of dry holes for seventeen years. But a wildcatter can't quit. He leased about 5,000 acres in the rolling, pine-covered hills of East Texas where most geologists scoffed at the idea of finding oil. One, however, Dr. A. D. Lloyd, shared Dad's belief. Dad sold enough interests to start drilling in 1927. His equipment was dilapidated. The bit jammed in the first well. He started a second. The pipe was hung up in the hole. In May 1929, he started the third. To keep going, he sold twenty-five dollar certificates for a 25/75,000th interest in the well and a 1/500th interest in the syndicate that owned 500 acres around the well. Waitresses, policemen, post office clerks, railroad workers—everyone bought. It was the midst of the Great Depression and maybe this was a way out. The well became a community project. Farmers and bankers would work on the well when Dad couldn't pay a crew. Inch by inch, the well reached 3,536 feet when it brought up a core with some oil-saturated sand.

When the word spread that Dad was going to drill in the well on the morning of October 1, 8,000 men, women, and children swarmed around the well. The cement plug was drilled through. Hour after hour, the bailer brought up only mud. At midnight most of the crowd was still there and bedded down in cars and wagons. They waited through the second day. Then the miracle occurred. There was a gurgling in the pipe and oil and gas sprayed over the top of the derrick. People laughed, shouted, wept, and bathed in oil.

The discovery was too much for Dad. His maze of financing resulted in so many lawsuits that he asked for voluntary receivership, then sold the well and all his leases to H. L. Hunt, of Dallas, for $1,500,000. Hunt, a professional gambler in Arkansas, had won an oil lease in a poker game. It produced. His gambler's instinct, as evidenced by buying Dad's snakepit of lawsuits, launched him on an oil exploration course which eventually made him one of the richest men in America.

No one really knew how big Dad's discovery was. Most of the major companies stayed out, enabling thousands of small operators to get leases. East Texas was the genesis of more oil fortunes and independent oil companies than any other field in history. Four months after the discovery, a second well came in at Kilgore, twelve miles north, for 22,000 barrels a day. Less than a month later another gusher at Longview, twelve miles farther north, came in. They were thought to be separate fields. Wallace Pratt, of Humble, was among the first to realize that, geologically, it was one huge field—a great stratigraphic trap formed by an ancient sea. Humble's massive leasing campaign netted it 16 percent of all the acreage in the fantastic field which was soon producing over an area of 211 square miles—43 miles long and from 3 to 12 miles wide, containing 5 billion barrels of oil.

Within ten months from its discovery, East Texas brought disaster to itself and the industry. Oil had been selling for $1.10 a barrel. By June 1931, with a thousand wells completed in East Texas, it plummeted to ten cents. A bowl of chili cost fifteen cents. Oklahoma City and Greater Seminole were also in full production. The nation, its industrial activities declining, could not absorb such a deluge of oil. The state governments of Texas and Oklahoma shut down the fields by martial law until oil companies and operators worked out plans to restrict production on an equitable basis and the price of oil was gradually restored.

A small oil field discovered in Oklahoma in 1930 would ultimately mean more than even the

"Dad" Columbus Marion Joiner, the seventy-year-old wildcatter, in white shirt and tie, being congratulated by his geologist, Dr. A. D. Lloyd, at the time of the 1930 discovery of the great East Texas oil field, below.

TEXAS MID-CONTINENT OIL & GAS ASSN.

Kilgore, Texas, in the heart of the East Texas oil field—another townsite drilling boom, below.

EXXON

MRS. E. L. DEGOLYER

Everette Lee DeGolyer (seated), father of petroleum geophysics, in Mexico,—1910, after locating the well that produced more oil than any single well in history.

treasure of East Texas, since the method of its discovery gave explorers a new tool to find billions of barrels undiscoverable any other way.

Everette Lee DeGolyer, father of this new method, was a renowned oil finder. In 1910, as a young University of Oklahoma geologist, he went to Mexico for a British company, Mexican Eagle, and discovered the famous Potrero del Llano No. 4, which eventually produced 130 million barrels, the greatest quantity any single well ever produced. Later, backed by Mexican Eagle, he founded Amerada Petroleum, which became one of the most successful oil exploration companies in the United States. DeGolyer's passion for exploring new oil-finding techniques led him to the successful development of *refraction* seismograph to locate buried salt domes. The method measured the time it took sound waves from a dynamite explosion to travel through rocks, to a recording instrument several miles away. Sound waves travel at different velocities through different types of rocks, traveling twice as fast through dense salt rock as through clays, shales, or sands. The method was a brilliant success in locating dozens of new Gulf Coast salt domes with oil production. Unfortunately, it wouldn't work any place else because of the Gulf Coast's simple geological conditions. There were no other high-speed rocks, such as there were in other areas, to complicate the readings.

DeGolyer abandoned the method and began researching *reflection* seismograph. Its principle is to generate sound waves by explosion of dynamite near the surface and record travel time to and from a buried reflecting rock bed to determine its depth. After shooting many points, depths are correlated. If the rock bed is arched or folded underground, the shape of the structure can be mapped to determine if it formed a trap.

Amerada drilled successive dry holes on "structures" delineated by reflection seismograph. DeGolyer was referred to, in oil circles, as "crazy with dynamite." He persisted in perfecting the technique. In 1930 he found a perfect diagnostic test—the center of the deepest hole in the Seminole plateau. Surface formations dipped into the hole from every direction, and it was ringed with dry holes. However, his reflection seismograph picture showed that far under the center of the hole the rock beds humped up into a little dome that might be an oil trap. When the drilling bit pierced the dome, the Edwards field began flowing 8,000 barrels of oil a day. Now, it was the whole industry that went "crazy with dynamite."

This was the most important well drilled in America since Spindletop. Reflection seismograph revolutionized prospecting as completely as Spindletop had done. Half of all the oil that would be discovered from that date on would be discovered by reflection seismograph in structures which would not have been found otherwise. Reflection seismograph made the 1930s an even more abundant oil-finding decade than the spectacular one just ended.

In the midst of their excitement over the seismograph method, explorers were given another major oil-finding tool. Unexpectedly, this one was the result of a catastrophe.

SHELL

This "gusher" was a different kind—earth, not oil. Such a one, as shown above, creating an artificial earthquake with dynamite, exploded the oil industry into the seismograph age in 1930. The industry now had its greatest oil finding tool!

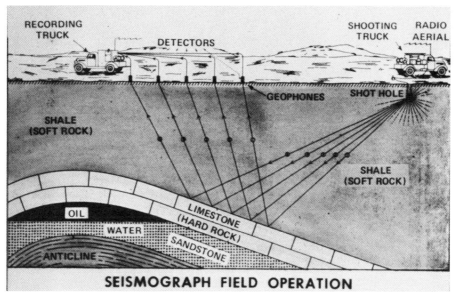

SEISMOGRAPH FIELD OPERATION

AMERICAN PETROLEUM INSTITUTE

The Conroe field, thirteen miles northwest of Houston, a new giant field, had been discovered in 1932 by George Strake, a Houston independent. He had convinced Wallace Pratt he had a good prospect, and Humble bought a half-interest to finance the discovery well and development. There was no forest of derricks—it was one of the first fields to be developed by orderly well spacing in joint agreement with all the lease owners, to avoid waste in overproduction. However, in January 1933, one of the wells blew out, catching fire. It was controlled, but a second well blew, the derrick collapsed and disappeared in a great hole in the earth. Like the crater of a volcano, the hole erupted, flowing 10,000 barrels a day.

Humble Oil carried out an incredible engineering operation to kill the crater. Two years before, a young Oklahoman, John Eastman, had perfected a device to intentionally drill crooked holes at Huntington Beach, California, to produce oil from part of a structure that lay under the ocean bottom. He guaranteed Humble he could drill a slanted well to get to the bottom of the cratered hole, pump water and mud down, and stop the oil flow. It took two months, but he hit the hole on target. After flowing 1,500,000 barrels of oil it stopped; the damage was repaired and the hole plugged. Drilling crooked holes now became an art to finding oil on salt dome flanks, under rivers and lakes, and wherever natural

Subsurface geology was revolutionized in 1932 when Shell Oil Company introduced in California and Venezuela the use of electric well logging invented by Conrad and Marcel Schlumberger of Paris (shown below). With it, geologists could tell types of rocks drilled through, whether they could be a reservoir for oil or gas and whether oil or gas was there.

SCHLUMBERGER

EXXON

Conroe, Texas, in 1933, above—a giant oil field that produced more than oil. One of the first fields to introduce well spacing to stop over-production, its great catastrophe resulted in a new oil-finding tool—directional drilling.

or man-made barriers to a likely prospect needed to be circumvented.

Flamboyant wildcatter, Ernest W. Marland, who discovered so many giant fields in Oklahoma, California, and Louisiana, was flattered to have J. P. Morgan become his banker and part-owner of his company in order to expand operations. However, during the depression he lost his company to the bankers. It was renamed Conoco. Marland was broke, but flung himself into politics and did more for the oil industry than he had as an oil finder. In 1934 he was elected Governor of Oklahoma and achieved two successes assuring him a place in state and oil history.

The Oklahoma City field had eventually surrounded the capitol grounds. Marland knew the wells would drain oil belonging to the state. His proposal to lease state lands was vigorously opposed by residents and those who thought it would spoil the capitol's beauty. A battle developed between city and state over authority to permit drilling. When the city council refused to permit leasing, Marland decreed the capitol grounds a military zone. He leased the land in open bidding and, surrounded by guards and journalists, Governor Marland drove the stake for the first well near the executive mansion. At his own expense he had landscaped the capitol grounds with exotic shrubbery and thought that derricks also were economically beautiful for the state.

His greatest achievement was formation of the Interstate Oil Compact uniting producing states in exercising control over conservation of petroleum resources by preventing physical waste in any form. Marland skillfully brought about agreements in 1935 from which grew an organization supported by twenty-one states. Never again would wasteful boom field practices dissipate the nation's oil resources. Participating states formulated and enforced regulations requiring good engineering practices and waste prevention from overproduction and overdrilling.

Oklahoma Governor Ernest W. Marland, center, the former great wildcatter, defies city authorities and, decreeing the capitol grounds a military zone in 1935, lines out the location for the first oil well on state capitol grounds, patrolled by state militia. In uniform, First Lt. Ross H. Routh, Executive Officer, Adjutant General's Office.

"Oil wells make mighty good landscaping"— according to Oklahoma Governor E. W. Marland, who forced drilling on state capitol grounds.

CITIES SERVICE

The art of petroleum engineering had been developing rapidly. Engineers learned that to prolong an oil field's life, oil and gas should be produced at a maximum efficient recovery rate, controlling the amount of gas produced to bring up oil instead of wide open production leaving millions of barrels unrecoverable. Voluntary agreements to share in orderly oil field development by well spacing would yield more oil for each lease owner than had earlier derrick-to-derrick madness. Gas could be reinjected into the formations to bring up more oil. Wells could be drilled to inject water to move oil to well holes so it could be pumped up.

Wasteful periods of flush production with resultant low field oil prices were eliminated, and recovery of the nation's oil and gas resources increased. Credit for these developments was due to the effective agency for self-regulation, created, supported, and administered by the oil states themselves and fathered by Ernest W. Marland who died, almost penniless, in 1941.

Prolific Texas Panhandle discoveries extending into Oklahoma and Kansas produced such enormous quantities of gas, as well as oil, that in 1930 the Natural Gas Pipeline Company of America, a consortium of companies, built the world's first high-pressure, thin-wall, large-diameter, cross-country gas transmission line. It made reasonably priced natural gas available to Midwestern cities. The Fritch, Texas, to Chicago line, a portion of which is shown at left, opened up transcontinental pipelining and inaugurated a new Natural Gas Age.

CITIES SERVICE

In 1933, Wallace Pratt negotiated for Humble the largest oil and gas lease ever purchased in the United States on the million-acre King Ranch in southwest Texas, shown at left. Today it has 5,000 oil and gas wells. Under Pratt's guidance, between 1930 and 1937, of twelve giant oil fields discovered in Texas, Humble owned either all or a great part of them, giving it the greatest reserves of any oil company in the nation.

EXXON

TEXACO INC.

The oil industry's first submersible drilling barge, Texaco's Gilliasso, right, went into service in 1933 at Lake Pelta in coastal Louisiana. These drilling barges made possible the development of oil and gas fields in Louisiana bayous and were the forerunners of the huge offshore drilling rigs yet to come.

70

TEXACO INC.

The depression launched the growth of extreme competitiveness among oil companies to capture the gasoline market. Companies vied to introduce and emphasize free services, left. The number of service stations jumped from 121,513 in 1929 to 241,358 in 1939. Gasoline sales increased from $1.8 billion in 1929 to $2.8 billion in 1939. There were 5,700,000 more automobiles, buses, and trucks on the road from 1930 to 1940. However, the phenomenal growth in petroleum product sales was for oil burners for home heating. In 1929 only 100,000 homes could afford the new, automatic oil burners. But by 1935, cheaper burners and cheaper fuel resulted in 1 million homes installing them and by 1940 there were 2 million homes with oil heat.

U. S. DEPT. OF LABOR

The link between oil and soil was one of the nation's great new strengths. Automobiles, trucks, and roads enabled farmers to expand production and get produce to market. In 1906 eleven companies began manufacturing the first gasoline-powered tractors which freed farmers from bondage to horses and mules. By 1930 there were 920,000 tractors on America's farms and 1,567,000 by 1940, such as one at right.

The refining industry was revolutionized for the second time in 1938. Dr. William Burton's thermal cracking process in 1913 had doubled the yield of gasoline from a barrel of oil. In the 1930s, a French engineer, Eugene Houdry, developed a process (research financed by Socony-Vacuum, now Mobil, and Sun Oil Co.,) which used a catalyst, rather than increased heat, to produce high-octane gasoline at quality and yield levels never before attained by conventional thermal means. The first commercial unit at Sun's Marcus Hook, Pennsylvania, refinery is shown at right. The Houdry catalytic cracking process made 100 octane gasoline possible. By the time World War II erupted, twelve U. S. plants with 132,000-barrel-a-day capacity became critically important sources of fighting grade gasoline for our Allies. Without it, the Royal Air Force (RAF) could not have won the Battle of Britain.

As the 1930s closed, oil and gas were supplying the United States with 44.5 percent of its total energy requirements as contrasted to 1920, when coal supplied more than 78 percent. But with war already started in Europe the American industry would face its greatest challenge.

SUN COMPANY

PHILLIPS PETROLEUM

Another phenomenal development of the 1930s was the use of Liquefied Petroleum Gas. LPG—or bottled gas—gave farms and rural communities the same advantages in their homes, tractors and stationary equipment as their city cousins had with their new piped natural gas to homes and factories. Sales of bottled gas jumped from 18 million gallons in 1929 to 313 million gallons in 1939. Left, LPG on the farm.

EXXON

This "Christmas Tree" as it is called in the oil industry, presented the American people with billions of barrels and trillions of cubic feet of gas that otherwise would have been lost. It symbolizes the growth of the art of petroleum engineering as its unique rigging of pipes and valves at the mouth of an oil well eliminates waste by keeping oil and gas production under control.

U. S. NAVY

1942—America's darkest war year. Tankers were America's and the Allies's lifeline, but the most vulnerable. Ninety-five percent of all oil and gasoline to the industrialized East Coast, which used 40 percent of all petroleum products in the nation, came by tanker from Gulf Coast oil fields and refineries. The "wolf pack" of Nazi submarines concentrated on preying on tankers. Sinkings exceeded by large tonnages the replacement rate. From February 1 to the end of May 1942, fifty tankers were sunk, such as the one above. Oil supply to the East Coast was virtually cut off, and the whole petroleum distribution system was disrupted.

WINNING WORLD WAR II:
"A Matter Of Oil, Bullets And Beans"—
Admiral Chester W. Nimitz.

Oil was the indispensable material in World War II. From beginning to end it was an oil war.

Oil and oil products constituted more than twice the combined tonnage of all other supplies shipped overseas, including men, weapons, ammunition, and food. Almost 7 billion barrels were consumed between December 1941 and August 1945, to meet the requirements of the United States and its Allies. Nearly 6 billion of this enormous total came from the United States.

Oil did more than fuel and lubricate ships, airplanes, and motorized ground equipment. In the field, it fueled kitchens, powered radios and telephones, kept hospitals going. From oil came the toluene for TNT that went into bombs, chemicals to produce synthetic rubber, asphalt for airfields, jellied gasoline for flame throwers, kerosene for smoke screens, and wax for packaging food and equipment. It went into medicines, anesthetics, insecticides. Oil powered the gigantic domestic war production program and met essential civilian requirements. In a thousand forms oil made the difference between winning or losing the war. Defeat of Germany and Japan stemmed from their lack of it.

The single most important ingredient in winning this worldwide oil war was the extraordinary, unprecedented partnership between the U. S. government and the American oil industry in the Petroleum Administration for War, under Secretary of Interior Harold L. Ickes. The motto of this government-industry team was: "The difficult we do immediately, the impossible takes a little longer." The miracles they achieved in production, transportation, shipbuilding, pipelining, refining, and distribution were incredible. At war's end, the Army-Navy Petroleum Board of the Joint Chiefs of Staff summed it up by saying that "at no time did the Services lack for oil in the proper quantities, in the proper kinds and at the proper places. . . . No Government agency and no branch of American industry achieved a prouder war record."

Initially, the partnership seemed unlikely. Hating Secretary Ickes was an occupational disease among oilmen who thought he wanted to be the "Oil Czar" of America and place the industry under federal control as part of the New Deal's recovery

"The Old Curmudgeon," Secretary of Interior Harold L. Ickes, below, who forged the great partnership between government and the oil industry, which was the single, most important ingredient in winning the war.

DEPT. OF THE INTERIOR

EXXON

From 1941 to 1943, the British Navy depended upon Venezuela for 90 percent of its needs. Increased American production and tanker building had not yet gotten into full swing. Venezuela had become the second largest producing country in the world, next to the United States. Its production from such fields as Lake Maracaibo, above, had been primarily developed by American oil companies since 1923. They controlled two-thirds of Venezuela's half-million barrels a day. Tankers from Venezuela to Europe were also a primary target of Nazi U-boats.

TEXACO INC.

program. However, in May 1941, when President Franklin D. Roosevelt, anticipating America's involvement in the European War, appointed Ickes the Petroleum Coordinator for National Defense, the Secretary stunned oilmen. He appointed Ralph K. Davies, Vice-President of Standard Oil Company of California, as Deputy Administrator, with authority coequal to his own. He called a meeting in Washington of a thousand of the most prominent major and independent oilmen and told them it was up to them to form a team. "The right of an industry such as yours to function in the American way is being threatened—not by your Government, but by totalitarian powers," he said. "The Old Curmudgeon," as Ickes was nicknamed, united government and industry in the job of defending America by turning it over to the men who knew how to do it best—oilmen.

When the nation went to war, the organization, renamed Petroleum Administration for War, was in place, staffed by oilmen from throughout the country—not bureaucrats. The most highly competitive industry in America became a single force, forgetting its animosities and rivalries, and pooling its know-how and trade secrets to win the war.

Oil was the lifeblood of the war, but pipelines were its arteries. Of all the remarkable wartime industry accomplishments, the saga of the pipeliners was unexcelled—in the United States, the Far East, and the battlefields of Europe and North Africa.

In the spring of 1941, the British lost 182 tankers to enemy action, so the United States loaned them 50, severely reducing tanker capacity

The war's most famous tanker was Texaco's S. S. Ohio, left. In 1942, it saved the Mediterranean island of Malta, whose bombers and submarines maintained the only effective striking force against Axis supply convoys to North Africa. Blockaded by the Axis, Malta only had a twelve-day supply of gasoline and food when the Ohio, carrying 13,000 tons of 100 octane gasoline, and a convoy of destroyers and supply ships made an attempt to reach it. Most of the convoy was destroyed, but the badly crippled Ohio made port, enabling the RAF to control the Mediterranean.

GULF OIL

Petroleum geophysics went to war to find Nazi U-boats. The use of a magnetometer, an instrument to register the intensity of magnetic attraction of buried rocks in order to locate possible oil traps, had been developed in the 1930s. It was particularly effective in locating buried granite mountain ranges along whose flanks might be oil and gas fields. Gulf Oil Corporation scientists developed the world's first airborne magnetometer in 1939. Mounted in a torpedo-shaped bird and towed behind an airplane, it recorded the continuous magnetic profile of its flight path. A week's flying could give the geophysicist basic data for a magnetic map covering several thousand square miles. However, before it was ever used to look for oil traps, it was pressed into wartime service to magnetically locate the presence of the vicious Nazi U-boat wolf pack preying on U. S. tankers.

The desperate tanker situation began to turn around by the end of 1943 when new tanker construction got into stride. Two tankers every three days, such as the one at right, slid down the ways in 1943 and 1944. The Allies lost 550 tankers, but built 908, of which the United States constructed 716. On VE-Day the United States was operating 800 tankers with capacity equal to the entire prewar world petroleum fleet.

to supply the East Coast. In June 1941, eight major oil companies recommended building a twenty-four-inch crude line from Texas to the East Coast. Their applications for priority steel were denied. However, at their own expense, they surveyed the route and did design engineering. In the dark days of 1942's massive tanker sinkings, the desperate need for a pipeline was obvious. With priorities granted, the industry was ready for the most spectacular pipelining job ever attempted—moving five times as much oil in one pipe as ever had been moved before, moving it halfway across the continent, and building it faster than any pipeline had ever been built.

In August 1942, under the supervision of Burt Hull, President of Texaco's Pipeline Company and dean of pipeline builders, thousands of pipeliners began working on all sections of the 1,254-mile line at once, battling swamps, rivers, floods, mountains, rocks, snow, and shortage of manpower. In a record-smashing 350 days the Big Inch was completed, and 300,000 barrels of crude oil were reaching eastern refineries daily.

Four months before the Big Inch was finished, construction began on a 1,475-mile parallel line—the Little Big Inch—to feed 235,000 barrels daily of refined products—gasoline, diesel, fuel oil—from

PETROLEUM ADMINISTRATION FOR WAR

PETROLEUM ADMINISTRATION FOR WAR

When it hit the Alleghenies, the incredible Big Inch crude oil pipeline from Texas to the East Coast, had to climb almost vertically in some places as shown above.

Texas refineries to the East Coast. The pipeliners beat their own record by completing the line in 225 days.

Together, the Big Inch and the Little Big Inch moved a total of 368 million barrels of oil and oil products to the East Coast by the war's end. To match this would have required a pool of about 40,000 tank cars, operating daily, that simply did not

U. S. SIGNAL CORPS

Lightweight mobile military pipe, above, being flown to China to keep its petroleum supply lifeline open.

exist. The heroic pipeliners were crucial to victory.

The development of a simplified, lightweight, military mobile pipeline by Sidney S. Smith, a Shell Oil Company pipeline engineer, played an indispensable role in the combat zones. Smith perfected a four-inch, lightweight pipe connected by special couplings that eliminated need of welding and could be quickly moved by inexperienced men. The pipelines were totally salvable, could be used over and over again in successive campaigns, were less vulnerable to bombing, and eliminated the necessity of trucks and extra manpower. Portable military pipelines went everywhere with the Allied armies in Europe and North Africa. The pipeliners were often minutes behind the retreating enemy. On one occasion in France they were energetically staking a route across a field when a soldier dashed out, panting, "The General tells you to get your men the hell out of here. The infantry hasn't taken this field yet!"

In the Pacific, military pipelines provided a quick and continuous flow of oil from tankers to troops well inland. The most remarkable military pipeline achievement was the laying of the 1,800-

TNT for bombs came from crash building of petroleum toluene plants such as Indiana Standard's plant, shown above.

From a test tube and a billion dollar crash program, the petroleum industry provided ingredients for synthetic rubber from such butadiene plants as the one above at Baytown, Texas. Without them the war could have been lost.

mile line across India and Burma into China, crossing deserts, jungles, gorges, and mountain ranges.

Every bomb, shell, and torpedo that the Army and Navy hurled against the enemy was loaded with TNT, whose basic ingredient is toluene, a clear aromatic liquid. During World War I it was a byproduct of coke from coal. In 1939, total annual U.S. toluene production was 25 million gallons, all from coal. But in 1940, American refiners, perfecting high-octane gasoline, discovered they could produce toluene from oil. TNT requirements for World War II were ten times those of the First World War. Each 500-pound bomb required 40 gallons of toluene. The larger bombing raids over Germany used 10,000 gallons of toluene per minute. The oil industry built dozens of toluene plants. In 1944 the nation produced 208 million gallons of toluene, of which 81 percent was from petroleum.

Next to oil, rubber was the most critical raw material in the war. Without it neither the military machine nor the civilian economy could operate.

At right, synthetic rubber, made from oil, coming off a conveyor from extruder, is then rolled to be made into products.

The most spectacular achievement of wartime refining was the billion dollar crash production program of 100 octane gasoline, boosting production from 40,000 barrels daily in 1941 to a peak of 514,000 barrels daily in 1944. Also, quality was vastly improved giving aircraft more speed, power, quicker take-off, longer range, and greater maneuverability. Quantity and quality meant the victory margin for air superiority. At left is the world's first hydrofluoric acid alkylation unit for 100 octane gasoline built in 1942 to use a process developed by Phillips Petroleum Company. The process was used in more than thirty plants and was responsible for a large part of the Allied forces' high-octane gasoline.

PETROLEUM ADMINISTRATION FOR WAR   PHILLIPS PETROLEUM

"Our secret weapon" is what Fleet Admiral Nimitz called the technique developed for refueling at sea. Never before in the history of naval warfare had a huge fighting force been independent of its bases. This made a sustained offensive possible at all times. At left, an airplane carrier simultaneously takes fuel oil and aviation gasoline from a tanker. To fill the fuel tank of one battleship took enough oil to heat a home for more than 500 years. Tankers became so skillful at refueling at sea that they could fuel a mammoth aircraft carrier on one side and two destroyers on the other while all four were steaming along at fifteen knots.

To keep the U. S. Air Force operating a single day worldwide required fourteen times as much gasoline as was shipped to Europe for all purposes in the First World War. Above, the wing-tip fuel tanks of fighter planes are being refueled.

U. S. SIGNAL CORPS

The prolonged, bloody struggle for the Pacific island of Guadacanal, which was won in February 1943, was a turning point in the war. Here, on land, American soldiers learned that the supposedly invincible Japanese jungle fighters who had overrun half of Eastern Asia could be beaten. The Navy learned how to fight night battles and shoot down enemy planes. After Guadacanal was secured, the Navy lost no more battles. As shown at left, on the beaches of Guadacanal it was the constant flow of barrels of gasoline and fuel oil that made the vital difference.

PETROLEUM ADMINISTRATION FOR WAR

The gasoline jerry can, alias "kraut can," and "blitz can," was a special hero to Allied fighters everywhere. They were ingenious, sturdy cans holding only four and a half gallons of gasoline each, but they were easy to stack and carry, convenient to fill and empty. They provided fuel for land equipment, radio and radar power, cooking, light, heat, and refrigeration. Their immediate availability was often the difference between victory and defeat. There were 13 million cans in use on the European western front alone. The jerry cans at right are being filled at the Fifth Army Gasoline Dump, Coverta Sector, Italy.

Paraffin wax, a petroleum by-product, became an important military "weapon." The small prewar production had been used mainly to coat paper milk containers, butter cartons, and other food packages. The war made it indispensable for packaging field rations for overseas and dip-coating munitions and weapons for combat zones to protect them from moisture. It was so important it was allocated early in the war, but by 1944 production was increased to 80,000 tons, of which more than 60 percent was for the military. The worker, at left, is peeling excess wax from the mold.

PETROLEUM ADMINISTRATION FOR WAR

PETROLEUM ADMINISTRATION FOR WAR

The world's biggest paving job happened when asphalt went to war. Worldwide war made asphalt—the bottom of the refining barrel—a priority product for building airfields and access roads to camps, repairing highways, constructing military highways in Pacific bases, and for roofing and waterproofing.
The Corps of Engineers and the Seabees went everywhere with bulldozers and drums of asphalt. Long before the Corps of the Engineers finished war construction, it was estimated they had put down enough paving to build fifteen 20-foot highways from New York to San Francisco. At right, asphalt is being laid at Agana Airfield, Guam.

GOODYEAR

One of the most unusual and important uses the Allies made of synthetic rubber during the war was to fool the Germans with a ghost army of make-believe rubber weapons—planes, tanks, trucks, and P. T. landing boats, developed by the U. S. Army Engineers. Deflated dummy planes, such as the one above, could be carried in a case no bigger than a golf bag. Hundreds of them could be positioned to trick the Germans into thinking the Allied air strike force was greater than it was. The "ghosts" were so easy to handle that twelve men could inflate 360 Sherman tanks in half an hour and carry them into positions where they drew some of the enemy's fire away from real tanks. Later, the trick targets could be deflated, loaded onto a few trucks and set up again on the next battlefield. German artillery shelled "ghosts" mercilessly. Prior to D-Day, concentrations of ghost weapons were stationed at various places in England to confuse the Germans as to where the real invasion would originate. During invasions, dummy landing craft drew German fire relieving the pressure on real landing craft loaded with troops.

SHELL SOCAL

With the men away fighting on the world's battlefronts, solving technological problems, laying pipelines, building refineries and petrochemical plants, American women went to war to help solve the "manpower" shortage as shown above. They worked in refineries, service stations, factories, and gave "yeowoman" service.

Rubber was essential for tires, electrical equipment, parts for planes, tanks, jeeps, battleships, gas masks; and hundreds of other vital uses. The United States, the world's greatest rubber-consuming nation, had no supplies of its own. When the Japanese overran 90 percent of the world's greatest rubber-growing regions in the Far East in the spring of 1942, the threat of a rubber famine could have meant total disaster for the Allies. However, the petroleum industry already was gearing up to provide a chemical miracle to produce synthetic rubber. In the race to produce 100 octane gasoline, petroleum scientists had discovered a by-product—butadiene—which could turn oil into rubber. A billion dollar crash program, beginning in 1942, produced enough butadiene to make more than 900,000 tons of rubber annually, more than half again as much as the total consumption of natural rubber in the United States in the largest prewar year. A major new industry was born, and the Armed Services never lacked for rubber.

At home, in addition to herculean accomplishments in transportation and miracles of molecules, the oil industry drilled 13,400 wildcat wells, more than in any similar period in its history. From 1941 to 1945 oil production increased by 27 percent and refinery runs by 30 percent. Although rationing of civilian uses of gasoline and fuel oil reduced consumption greatly, what kept the nation going was the dramatic expansion of the natural gas industry.

Natural gas production increased by 55 percent, providing the equivalent of more than 2 billion barrels of oil—more than one-third of the entire wartime domestic oil production. Total liquid gasoline production from natural gas, which was 150,000 barrels daily at war's beginning, was doubled.

Natural gas provided industrial and domestic fuel. Its products were essential for production of 100 octane gasoline and synthetic rubber. It supplied carbon black for synthetic rubber tires and hydrogen for more than half the synthetic ammonia used for explosives.

Never before had a nation so effectively mobilized its natural and human resources to be used on such a global scale. On August 14, 1945, when Japan surrendered, oil men immediately began to "roll up the barbed wire." Within six months the Petroleum Administration for War was out of business—the great industry-government partnership dissolved. Oil men eagerly went back to competing with each other and to tackling a new problem of the magnitude of those they solved during the war— replacement of the enormous wartime drain on the nation's oil and gas reserves.

The success of World War II was essentially the story of American oil production, know-how, crash programs, and coordination of worldwide efforts. In 1941, the United States was producing 3,840,000 barrels daily whereas foreign oil production amounted to only 1,000,000 barrels daily. Nevertheless, global war could not have been fought on the scale necessary without all of the oil and products that the combined efforts of the Allies could make available. With the expertise and organization of the U. S. Petroleum Administration for War, foreign companies and nations pooled their resources, plants, transportation facilities, and knowledge to form a closely knit unit, capable of supplying oil and products for Allied military and civilian requirements everywhere.

The foreign areas of greatest importance were the Caribbean—Venezuela, Colombia, Trinidad— and the Middle East—Iran, Bahrein, Saudi Arabia. During the war the Allies managed to increase foreign production by 61 percent and foreign refinery runs by 40 percent. Initially, increases came from the Caribbean. Until the German and Italian troops were driven out of Africa, it was impossible to take advantage of the production and refining capacity of the Middle East.

The most important analysis of the foreign world oil situation was the report of the United States Technical Oil Mission, sent in 1943 to study the prospects of the Middle East. It was headed by

ARAMCO

Dhahran, Saudi Arabia, 1938, above, when the seventh wildcat discovered commercial oil. Limestone hills on Damman Dome helped Arabian American Oil Company geologists define the structure.

E. L. DeGolyer, father of petroleum geophysics. Before the war, the British Anglo Iranian Oil Co. had made Iran the third largest foreign oil-producing country, based on discoveries dating back to 1907. However, Standard Oil Company of California (Socal) had obtained a concession on the small island of Bahrein in the Arabian Gulf and discovered oil in 1932. Its success persuaded King Ibn Saud, of neighboring Saudi Arabia, to give Socal an exclusive concession on all of eastern Saudi Arabia for sixty-six years. Saudi Arabia was so isolated from the so-called civilized world that the first American geologists arrived wearing beards and Arab dress in order to avoid attracting attention. Socal and its partner, Texaco, discovered oil in 1938. When the war broke

ARAMCO

"We have given you abundance"—The Koran. 'Abd Allah as-Sulayman, Saudi Arabian Finance Minister, and Lloyd Hamilton, Socal representative, sign the sixty-six-year concession agreement in 1933 to explore for oil.

out in 1939, Saudi Arabia was producing only 19,000 barrels daily and Bahrein 20,000 barrels, but Bahrein had a 35,000 barrel-a-day capacity refinery. This oil was so strategically located and urgently needed that the U.S. government made scarce material available to double Saudi Arabia's production and increase Bahrein's refining capacity to handle both countries' oil.

The startling news that the DeGolyer mission brought back from its survey was that:

> "The center of gravity of world oil production is shifting from the Gulf-Caribbean areas to the Middle East, to the Persian Gulf Area, and is likely to continue to shift until it is firmly established in that area."

As the war that was fought on oil ended, no more prophetic words were ever written.

GULF OIL

Exploration in Kuwait, on the Arabian peninsula, began in 1936 by Kuwait Oil Company, jointly owned by Gulf Oil Company and Anglo Iranian. Commercial oil was struck in 1938, but in early 1942 the first few wells were plugged with cement as wartime priorities were to develop refining capacity in Bahrein and Saudi Arabian production. Not until after the war did development begin on what would prove to be the 84-billion-barrel Burgan field— the world's biggest.

87

PHILLIPS PETROLEUM

Output of American agriculture and a higher standard of living on farms improved dramatically in the decade after World War II due to the doubling of gasoline-powered tractors and the tripling of use of LPG. LPG not only gave farm families the heating, cooking, and lighting amenities provided by electricity in the cities, but it was being put to a variety of important new agricultural uses. Above, LPG powers a flame-weeding tractor. Below, LPG is being used in a barn to cure tobacco.

PHILLIPS PETROLEUM

ENERGY REVOLUTION: 1945-1970

Alarmed predictions that "we're running out of oil" after World War II's drain on American reserves, were as vociferous as those following World War I. Neither government nor industry were prepared for what actually happened. The decade, 1945-1954, marked the most phenomenal growth in energy consumption and production in American history and a revolutionary change in the sources of the nation's energy and the places to search for it. It also introduced the amazing new world of petrochemicals, profoundly changing the American way of life.

Between 1945 and 1954, consumption of oil and gas nearly equaled the total amounts of those commodities used up until then since the first commercial well in 1859.

The decade's consumption explosion resulted from the remarkable increases of motor vehicles from 31 million to 58.5 million; gasoline sales of 24.4 billion gallons annually to 51.1 billion gallons; domestic oil burners from 2.5 million to 7.3 million; tractors from 2 million to 4.3 million; diesel railroad locomotives from 3,816 to 23,531; more than tripling of sales of Liquified Petroleum Gas; and more than doubling of marketed production of natural gas.

The decade marked the rapid decline of coal as the nation's major energy source. At the end of World War I coal provided three-fourths of America's energy, and oil and gas provided only 15 percent, with hydropower and fuel wood providing the rest. In 1945 oil and gas provided 41.2 percent of total energy, but by 1954 they accounted for 61.1 percent.

After World War II natural gas became the "glamour girl" of the energy industry. This once unwanted and wantonly wasted resource more than doubled in its marketed production between 1945 and 1954 and captured 22 percent of the energy market.

Since the industry's beginning, natural gas had been used in cities close to producing fields, but until the 1930s a driller considered the discovery of a gas well worse than a dry hole as he had to cap it. Gas from oil wells was universally flared. However, the development of high-strength, thin-wall pipe and electric welding in the 1930s made it economical to lay long-distance transmission lines. Natural gas potentialities were beginning to be realized when war broke out, halting new pipeline construction. After the war, gas pipelining boomed. Gas compa-

Railroads were primarily responsible for developing the coal industry and the most important factor in its decline. After World War II diesel electric engines replaced coal-fired steam locomotives so quickly that by 1954 they accounted for 72 percent of all fuel costs for railroads.

EXXON

EXXON

To meet demands of increased consumption of oil products, refining capacity was increased by more than 50 percent from 1945 to 1954, from a total of 5 million barrels daily to 7.7 million. In 1949, Exxon's Baytown, Texas, refinery, at left, became the largest in the United States, refining 260,000 barrels daily. In 1982, it tied for first place with the Hess refinery in the Virgin Islands as the world's largest refinery. It has a capacity of 640,000 barrels daily out of the U. S. total of 18,700,000 barrels daily capacity.

To carry crude oil and products, the industry constructed another 25,000 miles of pipelines between 1945 and 1954, bringing the network total to 139,000 miles. At right, pipelines enter the huge manifold of the Midland Basin Pump Station in West Texas, which handles more than 200,000 barrels of crude daily as part of 1982's network of 174,519 miles of crude oil and products pipelines.

TEXACO INC.

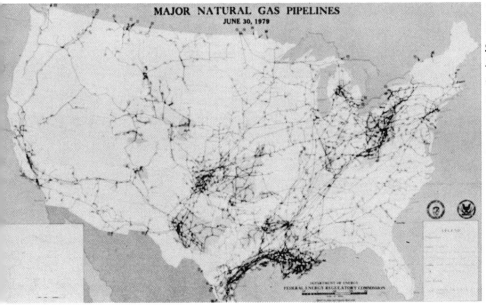

The natural gas boom after 1945 began the rapid increase of gas pipelines from 104,000 to 276,000 miles in 1982, representing an investment of $33.7 billion and fueling 40 million homes housing over half the population. The national network is shown at left.

DEPARTMENT OF ENERGY

nies purchased the war emergency Big and Little Inch oil lines. By 1947, Texas gas was moving to both the Atlantic and Pacific coasts. By 1957, every state in the union except Maine and Vermont was connected to the natural gas pipeline national network.

Congress had given the Federal Power Commission jurisdiction over pipeline rates in interstate commerce in 1938, but not over producers' prices. When gas prices advanced sharply, primarily due to general postwar inflation, there was a consumer revolt and pressure for federal regulation. It came in 1954 with a Supreme Court decision decreeing that the Federal Power Commission should regulate the price of gas sold by producers in interstate commerce. This bombshell was the root of the energy problems the nation would face in the 1970s. Under price regulation, gas demand skyrocketed. Its cheapness depressed the market for coal and fuel oil, starting a chain of events which led to a lopsided, unbalanced development of energy resources and growing dependence on foreign imports.

Freed from wartime price controls, America's major oil companies and 15,000 independent oil men began a tremendous drive to replenish reserves and meet astonishing new consumption demands. Between 1945 and 1954 they nearly doubled production from 1.7 million barrels daily to 2.3 million barrels. Huge discoveries were made in the Rocky Mountain regions. During this period they drilled a total of more than 377,000 wells, of which over 251,000 produced oil and gas. However, the dry hole count was 127,000, bringing the total of dry holes drilled since the 1859 Drake well to 390,000. Explorers knew that oil and gas are the world's most hazardous, expensive gambling game. However, they were constantly proving that the way to find them is by multiple effort with the incentive to do so. Unfortunately, the first worm in America's energy apple came in 1948 when the United States became a net oil importer. This was due to the fact that cheap, regulated natural gas had competitively replaced much domestic fuel oil. Imports of cheap, Middle Eastern crude and products began increasing steadily.

The most portentous worldwide postwar oil and gas event was the American development of technology to explore and produce offshore. Limited shallow offshore drilling in the United States began in California fields as early as the 1890s, and oil had been found and produced throughout the country in shallow lakes and rivers extending from land discoveries. However, the real birth of the world's huge offshore industry dates from the first offshore Gulf of Mexico lease sales held by the state of Louisiana on August 14, 1945, the day before World War II ended. The first successful offshore

TEXACO INC.

The birth of the worldwide offshore oil and gas industry was in the late 1940s in the Gulf of Mexico offshore Louisiana, as shown above.

Explorationists searched unsuccessfully for commercial oil in Canada from the time of the 1859 Drake well until, in 1947, Exxon geologists located the LeDuc discovery well, below, in western Alberta, for its Canadian affiliate, Imperial Oil of Canada. This giant 360-million-barrel oil field spawned the Canadian oil industry, and during the following decade the region's enormous potential unfolded as one of the world's major oil and gas regions. Pipelines were built to the Pacific Coast and the Great Lakes region, so that instead of exporting crude to Canada, the United States imported from it.

EXXON

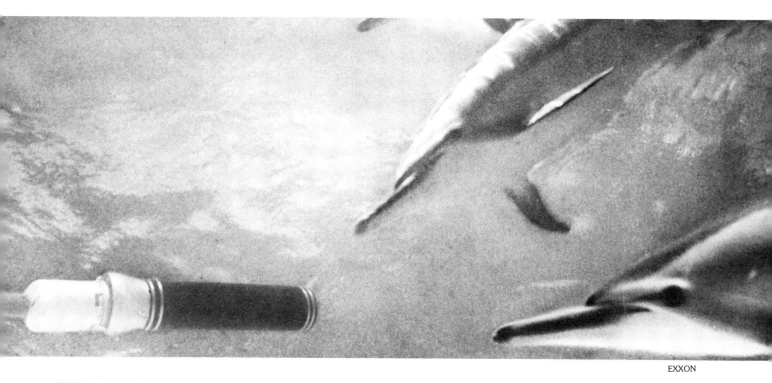

EXXON

"The Popper." Initially, dynamite was used to create shock waves underwater. Carefully handled, it did little or no harm to aquatic life. But fish were sometimes scared, fleeing to quieter waters, and fishermen objected. Consequently, Exxon invented a dynamite substitute. A mixture of propane and oxygen is ignited by a spark plug inside a rubber sleeve. The device, shown above, doesn't go bang like dynamite. It simply goes pop. It neither frightens nor hurts fish, and it costs only a penny a pop instead of fifteen dollars for a dynamite shot. A crew using a popper can do an underwater survey six times faster than by using dynamite and can work night and day in rough weather.

production in the Gulf came in 1947. By 1958, the industry had invested $3 billion in Gulf Coast leases and operations and had produced 200 million barrels of oil and 575 billion cubic feet of gas.

Major companies shifted to offshore exploration to look for big fields. By the mid 1950s, discoveries of onshore giants were becoming scarce. Offshore poker was too high-stake for the majority of independents. Not only were drilling costs much greater, but obtaining state and federal leases involved huge bonuses. Between 1958 and 1967, bonuses were about $2.5 billion more than normal expenditures for leases. This meant the industry had that much less exploration capital to put into the ground. By 1970 there were approximately 12,500 offshore wells, primarily in the Gulf of Mexico, producing about 15 percent of the nation's oil and 10 percent of its gas. This development cost $13 billion, including $6 billion in bonus, rental, and royalty payments to state and federal governments. The payout figure of a successful offshore discovery was estimated to take a minimum of ten years.

CITIES SERVICE

Laniscot I—*the pioneer seismic offshore exploration boat for the United States and the world. Its name incorporated the initials of the oil company group participating in its Gulf of Mexico surveys—Cities Service, Atlantic, Tidewater, and Conoco. It made its first seismic shot in 1946, launching a new era of seismic offshore exploration to find and develop a major portion of the world's potential oil and gas resources.*

Special water techniques could cover a greater area in shorter time than on land. As shown at right, sound waves are generated to travel deep below the seabed. These waves reflect from rock formations to the surface where detecting devices record them. By measuring time intervals required for impulses to travel down and back and interpreting the data, a geophysicist can determine general configuration of formations under the seabed and can identify possible oil and gas traps.

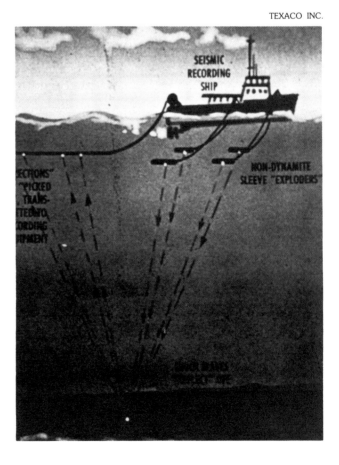

TEXACO INC.

AMERICAN HERITAGE CENTER, UNIVERSITY OF WYOMING

The potential of offshore exploration was revolutionized in 1956 when the world's first drill ship, Cuss I, shown above, drilled a well in 200 feet of water off the California coast. Its name came from the initials of the oil companies—Conoco, Union, Shell, and Superior—which had developed the floating vessel that could be anchored and could drill an average 5,000-foot exploratory hole in ten days. Eventually, the ship was to drill a total of 300,000 feet of holes, including one beneath 1,500 feet of water. In order to keep an eye on bottom operations in water too deep for divers, Cuss I had an underwater television camera which relayed pictures to a TV screen in the pilothouse.

ZAPATA

ZAPATA

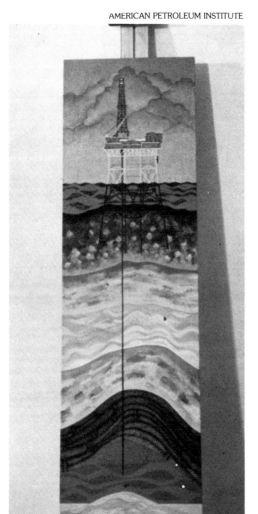

AMERICAN PETROLEUM INSTITUTE

Another 1956 giant breakthrough in offshore technology was development of the world's first jackup rig, The SCORPION, by Zapata Offshore Co., shown, left above, before drilling rig was attached. Jackup rigs are towed to a location where the legs are lowered and the platform jacked up electrically to a safe level above the sea. Maximum water depth in which jackup rigs are used is about 300 feet. At right above, is Zapata Offshore Company President and cofounder, George Bush, with his son, George, Jr., at the Scorpion commissioning ceremony. In 1981 Bush became Vice-President of the United States.

At left, an earth cross section showing undersea layers beneath an offshore drilling rig.

Sohio's petrochemical superstar, acrylonitrite, makes superior plastics for auto grills, (shown above), furniture, toys, and sporting goods because of its high-impact strength and outstanding ability to take colors.

Following World War II, rapid development of new chemicals from oil and gas spawned products that literally changed the world's manner of living—synthetic fibers, plastics, fertilizers, insecticides, and detergents.

In the United States, petrochemicals produce 75 percent of all fibers woven or knit into fabrics, virtually all plastics, half the plant fertilizers, and most of the rubber. There are more than 10,000 petrochemical products with hundreds of new ones added every year, providing substitutes for natural materials such as metals, wood, leather, paper, cotton, wool, and silk. Petrochemicals also create new materials such as foams and films.

Chemically, oil and gas are the most versatile raw materials. They are hydrocarbons containing a mixture of chemical compounds. Initially, industry took crude oil and *refined* out of it various chemical compounds already in it such as kerosene, lubricating oil, fuel oil, and gasoline. In the 1920s, American oil refiners and chemical companies began taking carbon and hydrogen atoms of petroleum and natural gas liquids and rearranging them into molecules to produce synthetic chemicals such as ammonia, ethylene glycol for dynamite and automobile antifreeze, and formaldehyde. Wartime crash development of synthetic rubber inaugurated the explosive postwar growth of the multibillion dollar petrochemical industry.

Just before the war, DuPont, after ten years of research, produced the first wholly man-made fiber—nylon. It was the forerunner of many synthetic fibers such as Dacron polyester and Lycra Spandex. Perhaps the most important man-made fiber discovery since nylon is DuPont's Kevlar, which is three times as stiff as fiberglass and five

New materials inspire new art forms, as demonstrated by the abstract sculpture, left, made from petroleum-derived polyurethane plastic.

ALLIED CHEMICAL

Oil and gas are the raw materials for synthetic fibers that provide 75 percent of all the fabric made in the United States. Below, researchers at DuPont, the pioneer in synthetic fibers, check materials for styling and wearability.

CONOCO

MOBIL

Clear plastic films from petrochemicals revolutionized food packaging, as typified by bread wrappers, above. Plastic films also are used for electrical and electronic insulation, magnetic recording tape, and as a base for photographic film.

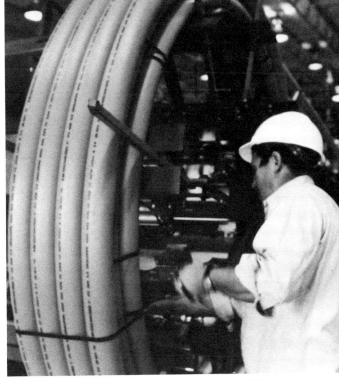
PHILLIPS PETROLEUM

Large-diameter plastic pipes, used extensively for water and sewer systems, and natural gas distribution, are shown above at Phillips Petroleum manufacturing plant in South Carolina.

times stronger than steel, pound for pound. Its principal use is in tire cords. It also is woven into cables strong enough to moor floating offshore oil rigs. Layers of Kevlar fabric will stop bullets from most handguns. Ballistic vests reinforced with the fiber protect a quarter-million police officers in the United States.

A chemical superstar was born in 1960 when Sohio researchers produced acrylonitrite, one of the world's major basic chemicals for producing synthetic fibers and plastics. Acrylic fibers include Acrilan carpeting and Orlon sweaters. Certain cotton and wool textiles contain as much as one-third acrylon by weight to improve resistance to shrinkage, wrinkling, and mildew, and to add strength. Acrylonitrile-butadiene-styrene, or ABS, plastics are superior for products where toughness and rigidity are required for broad temperature ranges and where hardness and stability must not vary appreciably. Consequently, they are used in such products as refrigerator linings, bodies for snowmobiles, travel trailers, and powerboats. ABS and other plastic resins have enabled Detroit to replace expensive, heavy-metal cars with more compact, lightweight, fuel-saving cars. A new car typically contains 200 pounds of plastic.

Union Carbide's development of vinyl plastics resulted in a host of applications such as phonograph records, raincoats, sheeting, wire and cable coating, and inflatable toys.

American petrochemical technology has been introduced worldwide. From the early 1970s to 1982, petrochemical plants have been built in seventy-three countries. Despite the enormous importance of the petrochemical industry to the world, it uses a surprisingly small amount of oil and

STANDARD OIL (INDIANA)

A Georgia farmer guides a cotton harvester through a field, above, that produced about 40 percent more cotton than the state's average. He had used new oil industry technology of applying needed micronutrients—magnesium, zinc, and boron—mixed with liquid petrochemical fertilizer to keep his soil in nutrient balance.

UNION

The mountain of sulphur, above, is turning a petroleum by-product into farm products. Sulphur, refined from oil, is converted by Union Oil Company in California to "popcorn sulphur"—little balls which are used to convert alkaline soil into good farming land.

gas resources. In the United States, during the 1970s and 1980s the amount of crude oil and natural gas used to make petrochemicals amounted to only about 4 percent of the total amount of oil and gas consumed for all uses.

One of the greatest contributions the petrochemical industry has made has been the vast economic growth in industrialized and developing countries in creating factories and jobs to manufacture the myriad products petrochemicals make possible.

Oil and gas had made America the world's greatest agricultural country. In 1900 one farmer produced enough to feed himself and nine other people. By the 1980s, with 6 million farmers—4 percent of the population—each one was providing enough food for fifty-six people, about a dozen of them living in other countries. America's food output averages about 30 percent annually more than its needs, and it is the nation's largest export commodity. By 1981, the United States was selling about $46 billion abroad annually. This world phenom-

Oil not only helps boost food production, but a completely natural, high-protein food is now made from oil—Tortula yeast—to help solve the world's nutritional problems. In the 1970s, Standard Oil Company of Indiana developed technology to make tortula yeast from ethyl alcohol derived from refined petroleum. The yeast, which is 53 percent high-quality protein, is dried into a product, resembling tan wheat flour, which can be added to virtually any food eaten anywhere in the world. It enhances flavor, color, and texture and is a much needed source of protein to combat malnutrition. Dozens of U. S. companies use it to enrich food products, and the technology is being exported to developing countries. At left, Amoco research team members are shown dining on tortula yeast-enriched foods—stuffed peppers, Swedish meat balls, pizza, a beverage, rolls, plus a tossed salad and desserts with dressings and toppings nutritionally enhanced by the oil-derived yeast.

STANDARD OIL (INDIANA)

enon is due primarily to four factors—farm mechanization with oil- and natural-gas-powered vehicles; petrochemicals for farm use, fertilizers, insecticides, and herbicides; development of high-yield hybrid seeds; and the private enterprise system where profit is the incentive for productivity.

In the 1930s the average yield of corn per acre ranged from 20 to 30 bushels. By 1976, with the use of high-yield hybrid seeds requiring heavy fertilizing—half of which comes from petrochemicals—per acre yields were well over 100 bushels and approaching the 200-bushel level.

The latest development to increase crop yields is analyzing soil to discover its deficiencies in any of the seven basic micronutrients required for plant growth—iron, zinc, manganese, copper, boron, molybdenum, and chlorine—and adding them to liquid petrochemical fertilizers. Pioneered by Amoco Oil Company, micronutrient application produces dramatic results. Such crops as corn, soybeans, sorghum, vegetables, pecans, cotton, sweet potatoes, and blueberries have increased 10 to 30 percent in volume, with improved protein quality.

To plant crops, today's farmers generally pull four separate machines over the same soil, consuming energy with every step. However, Phillips Petroleum, working with the University of Idaho, created a new planting process, called Ecofallow, shown at right, which tills, fertilizes, plants, and replaces soil in a single step. Fields that used to take weeks to prepare and plant can now be completed in a few days, cutting the farmer's energy requirements in half.

PHILLIPS PETROLEUM

SHELL

The introduction of petrochemical insecticides was mainly responsible for increasing crop yields by 50 percent between 1945 and 1965, as demonstrated, left, by spraying grapevines. Insecticides increased crop quality, efficiency of crop production levels. Petrochemical herbicides increased yields by 10 to 15 percent. Some petrochemical insecticides have been banned in the United States because of concern over their adverse effects on animals and humans in the food chain. However, petrochemical researchers have been devising new types which are nontoxic.

Invention of petrochemical foams which could be made into detergents revolutionized the soap business, providing a better product to wash clothes, dishes, and make shampoos and fine fabric cleaners. However, in the late 1950s petrochemical foams had created a serious environmental problem. They continued to foam when flushed down drains into water treatment plants. Municipalities complained about sudsing along local streams and creeks. Conoco chemical researchers solved the problem in the early 1960s by creating the first biodegradable detergents that would decompose in sewage treatment plants. At left, researchers in Conoco's Ponca City, Oklahoma, laboratory test new biodegradable detergents.

CONOCO

SOCAL

The building, right, in greater Los Angeles, is not the office building it seems to be. It is Standard Oil of California's Packard Drill site, sitting on top of the Crescent Heights Field, part of the oil-rich Los Angeles Basin. Constructed in 1966 at a cost of $4.9 million for equipment and building, all signs of petroleum operation are completely masked. Development drilling began in 1967, and production peaked at 25,000 barrels per day in 1968. In 1982, it averaged about 4,500 barrels of oil per day and about 2 million cubic feet of natural gas.

UNION

PUZZLE: How many oil-drilling rigs and offshore production platforms are in the picture above?

About a mile offshore Long Beach, California, are four artificial islands with structures resembling apartment towers, waterfalls, free-form sculptures, and hundreds of palm trees and shrubs. These islands, one of which is shown above, account for about 50 percent of California's offshore production and have produced over 360 million barrels of oil from some 700 wells. The unique islands were built in 1966 by Thums Long Beach Company, costing more than $41 million. Thums is an acronym based on its equal ownership with five oil companies— Texaco, Humble, Union, Mobil, and Shell. Each island comprises about 10 acres, built in 25 to 40 feet of water.

Drilling rigs, which are 178 feet high, are sound-proofed and camouflaged to look like modern high-rise buildings. People call up asking to rent apartments. Wells are drilled directionally from the islands, and rigs are moved along rails between drilling locations. Once an elderly lady excitedly telephoned authorities to report that "one of your hotels is moving." Before the islands were built, Thums hired a sound expert from a Hollywood movie studio to monitor noise levels along the shoreline to ensure the islands would not increase noise levels. Equipment is powered by electricity to prevent noise pollution. Electric golf carts are used to transport personnel.

Thousands of visitors from all over the world visit the islands annually. They are an oil industry Disneyland. Thums has received twenty-one awards from city, state and national groups for environmental protection, beautification, accident prevention, and engineering achievement. The project has proved that oil in large quantities can be produced in harmony with recreation and living areas. It also typifies the oil industry's concern about environmental responsibilities long before environmentalists launched crusades.

Records of the National Audubon Society's annual Christmas Bird Count are significant concerning the effect of oil refining on ecology. Audubon Society bird-counting teams set up stations nationwide to see who can count the greatest number of bird species in the same daylight period. The purpose is to learn if any species is under dangerous stress. Freeport, Texas, looks like an environmentalist's nightmare with its chemical and petroleum plants. Birds obviously don't believe what they see. From 1956 to 1967, Freeport was among the top North American stations in the annual census; from 1968 to 1974 it was among the top five. In 1972 it

Migrating and wintering snow geese almost conceal a gas-processing plant at Avery, Louisiana, a long-time oil- and gas-producing area.

Whooping cranes have been saved from extinction by the cooperation of Conoco and the Fish and Wildlife Service in their winter home at Aransas Wildlife Refuge on the Texas coast. Oil and gas have been produced since the 1930s. Drilling operations are shut down when necessary during breeding time.

set the all-time record with 226 species sighted. Significantly, John James Audubon, the famed naturalist, first went to Galveston Bay, where Freeport is located, in 1837 to explore and record its large bird population.

Energy activities can actually improve the environment for birds. In 1978 Standard Oil Company of Indiana's Casper, Wyoming, refinery was given an Audubon Society award for turning nearby Soda Lake, which had a high alkali content, into a freshwater pond to receive treated refinery waste water. New vegetation sprang up and freshwater shrimp thrived providing food for wintering birds. The lake now raises more ducks and geese than any other Wyoming site.

Louisiana has always held first place in U.S. oyster production. In the 1940s, the oyster catch declined as coastal oil production increased. Oyster fishermen immediately blamed oilmen. State university and oil company scientists found that oil had no effect on oysters. The real villain was a fungus existing in saltwater. The oysters were saved by preventing saltwater intrusion into the estuaries. The world's largest oyster bed lies in the middle of the Louisiana oil field at Marsh Island.

Shrimp fishermen were even more apprehensive about oil than oyster fishermen. Morgan City became shrimp capital of the world when jumbo shrimp were discovered in 1933. The city was the first staging area for the Gulf offshore oil industry. Shrimpers were openly antagonistic to the oil invasion. However, as shrimp production increased as rapidly as oil, they paid the oil industry their greatest tribute. Every Labor Day weekend one of the nation's most colorful festivals and pageants was held to celebrate and bless the shrimp harvest and the fleet. Beginning in 1967, Morgan City changed the name of the annual affair to the "Louisiana Shrimp and Petroleum Festival." Priests now bless the oil as well as the shrimp harvests.

There are more than 24,000 oil and gas wells producing in Gulf Coastal marshes and 14,000 outside the three-mile limit. In all other U.S. fishing grounds, except Hawaii, fishing has declined over the last quarter-century, but the Gulf doubled fish production in the 1970s. Thousands of offshore platforms create artificial reefs which attract a tremendous growing fish population.

Priests bless the shrimp and oil harvests at the Morgan City, Louisiana, annual "Louisiana Shrimp and Petroleum Festival."

UNION OIL

The world's largest jack-o-lantern—"Smiling Jack"—has been delighting the children and adults of Los Angeles since the 1950s, above. The huge smiling face is painted every Halloween week on an orange-colored gas storage tank at Union Oil Company's refinery, towering above the city. Its eyes and nose are eighteen feet high, its mouth seventy-three feet long and its teeth four feet square. Day and night it can be seen for miles.

Shell Oil Company has a sense of fun, too. It has whimsically decorated its horsehead oil pumps for oil wells in California as shown below.

SHELL

The exploration circle came full round when kerosene—the need for which prompted drilling of America's first commercial oil well in 1859—launched America's first manned moon-landing mission. At left, Apollo 11, riding a pillar of flame powered by kerosene, rises to clear its mobile launcher at Kennedy Space Center, Florida, on July 16, 1969.

When the lunar module, "Eagle," landed on the moon July 20, Astronaut Neil A. Armstrong took man's first step on lunar soil, as shown right, saying "That's one small step for man, one giant leap for mankind." The space suits of Armstrong and his companion, Edwin A. Aldrin, Jr., were made of layers of different petrochemical synthetic fibers. Parts of the rockets and space capsules contained many petrochemical products including plastics, synthetic fibers, and adhesives.

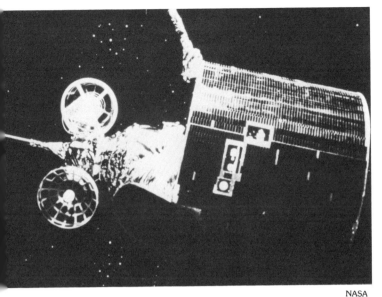

From doodlebugging to satellite. The oil industry's contributions to powering and providing materials for the space age have been "bread cast into the skies." The space programs' workhorses are the hundreds of satellites, such as the one at left, girdling the globe at intervals of 90 minutes or so. They take "pictures" of earth electronically by cameras, or sensor systems, which give geologists new clues of where to look for oil, gas, and minerals. Picture information from these satellites is also used in agriculture, forestry, water resources, weather studies, and by military and intelligence agencies.

NASA

NASA

The landsat, or land satellite picture, at right, shows a geologically complicated area in northeastern Utah taken at an altitude of 560 miles. Landsat pictures let geologists study larger surface areas than they've ever been able to see in one image, and mosaics of ever-larger areas can be put together. Landsat's electronic cameras are sensitive to varying brightness levels which can be translated by computers into color and black-and-white images. The sophisticated data make geologically interesting features stand out and are proving to be an increasingly valuable, new exploration tool for potential oil and gas areas.

ATLANTIC RICHFIELD

Prudhoe Bay State No 1—discovery well for America's biggest oil field on Alaska's North Slope in 1968, above. Drilled to 12,000 feet, it produced 2,400 barrels daily, opening up a field two to three times as big as the fabled East Texas. Enclosed in a metal building so that it can be serviced in severe weather, it is a monument to the development of American technology to explore and conquer one of the world's most hostile environments—the Arctic.

THE ENERGY CRISIS: 1970s

As the 1960s came to a close, national and international political developments were conspiring to end the abundance of cheap energy.

Oil and gas were now providing three-fourths of the nation's energy needs. However, domestic exploration by independent oilmen, who traditionally had found three-fourths of the nation's new oil and gas fields, had been reduced by almost half from 1956 following natural gas price controls. The number of independents had fallen from 20,000 to 10,000. Exploration costs had increased drastically, but crude oil prices had remained low. Demand for oil had doubled, and imports were supplying more than one-fourth of oil needs by 1970. Politicians and consumers alike ignored warnings from independents that the nation was headed for serious oil and gas shortages.

But in 1968 America flexed its oil muscles once again with the discovery of America's biggest oil field, Prudhoe Bay, on state lands on the North Slope of Alaska bordering the Arctic Ocean. An Arco-Exxon wildcat opened up a field containing a conservatively estimated 10 to 15 billion barrels of oil and 26 trillion cubic feet of gas, covering 250,000 acres. Other companies rushed to drill their leases, and a score of major companies promptly bid $900 million for additional state leases to explore one of the world's great new oil provinces.

Energy development in the United States was dealt a crippling blow as the result of an accident on an offshore exploration well on federal leases in the Santa Barbara channel off the coast of southern California in January 1969. A blowout on a Union Oil Company well drilling on a tract, jointly owned with Mobil, Texaco, and Gulf, spilled 10,000 barrels of oil into the channel before it was brought under control. It polluted beaches and created an oil slick over 200 square miles of ocean.

ALYESKA PIPELINE SERVICE COMPANY

In 1969 eight companies owning Prudhoe Bay reserves formed Alyeska Pipeline Service Company to engineer and build a pipeline with a 2-million-barrel-daily capacity, from the North Slope through 800 miles of Alaska's rugged terrain to the ice-free Pacific Ocean port of Valdez. They expected to complete this unprecedented engineering feat by 1973, and imported $300 million of pipe, whose first shipment to Valdez in September 1969 is shown at right. However, the pipe would lie idle in the storage yard for five years due to an oil disaster offshore California.

UNION

Two days before Union Oil Company's well blowout in the Santa Barbara channel in 1969, the famed resort beach was already a mess from storm debris, as shown above. The 10,000-barrel oil spill further fouled the beach and was acclaimed a national disaster by environmentalists. Four months after the blowout, after a massive cleanup, the beach was back to normal as shown below. There was no permanent damage to marine life.

UNION

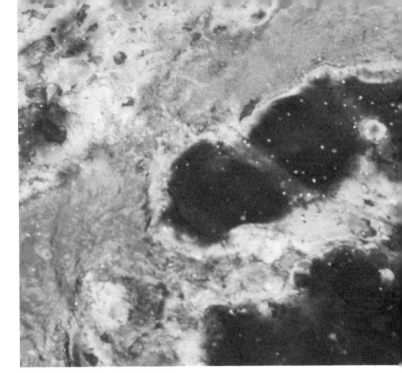

The biggest oil polluter in the Santa Barbara channel was not the oil companies, but nature. Fifty or more offshore oil and gas seeps, such as the one at Coal Oil Point at right, leak more oil into the channel each year than did the Union well blowout. Also, in the furor over the Santa Barbara oil spill little publicity was given to the fact that there were 16,000 U. S. offshore producing oil and gas wells and in twenty-five years of operation there had been only three other major oil spills in excess of 5,000 barrels.

ROBERT W. ESSER

The well had discovered a giant 221-million-barrel oil field, but it became a national media event presented as a major ecological disaster. The sight of dead birds, fouled beaches, and predictions of permanent damage to marine life inflamed public opinion against oil companies. The Secretary of Interior immediately suspended activity on thirty-five of sixty-nine Santa Barbara leases and ordered a leasing moratorium for all U.S. offshore areas. California banned new drilling on all state offshore leases.

The beaches were clean four months later, following a massive oil companies' campaign. Scientific studies over the following eighteen months showed complete recovery of the area environmentally, and the channel fish catch was greater than the year before the spill. However, the spill had created a new national disease—petrophobia.

Santa Barbara became the banner of an environmental crusade against pollution of all kinds—air, water, and land. It quickly exploded into a dominant political issue. By the end of the year, Congress had passed a hastily conceived National Environmental Protection Act which required all federal agencies to prepare a detailed statement of justification of "major federal actions significantly affecting the quality of the human environment." This was called an Environmental Impact Statement. In effect, it meant that interested citizens can challenge in the courts any federal agency's ruling on a matter affecting the environment on the grounds that the ruling does not comply with the intent of the Environmental Act. It gave environmental groups a powerful weapon to use in every area affecting energy development especially since 50 percent of the nation's remaining oil and gas potential, 40 percent of its coal, 50 percent of its oil shale, and 60 percent of its geothermal energy sources are located on federal land.

In April 1970, environmentalists fired their first shot under the new act. It stopped the building of the Alaska pipeline. The Wilderness Society, Friends of the Earth, and Environmental Defense Fund charged in court that the Interior Department had failed to make an adequate environmental impact review of the project. They claimed the pipeline, in traversing 400 miles of frozen tundra, three rugged mountain ranges, 70 streams and rivers, and one earthquake zone, would irreparably damage "the last great wilderness in the United States." They claimed it would adversely affect plant and animal life and result in disastrous oil spills caused by earthquakes and tanker accidents. A federal court

Two years before the Santa Barbara spill the American Petroleum Institute initiated a nationwide program to develop cooperative oil spill cleanup organizations, operating like a volunteer fire department. Oil companies, municipalities, state and federal agencies, and public organizations participate in 100 groups on the East, West, and Gulf coasts and on inland rivers and lakes. Above, an oil spill cooperative drills on the Connecticut River.

The largest cooperative is Clean Seas, Inc., founded by fifteen major oil companies after the Santa Barbara spill. It monitors 250 miles of southern California coastline. Below, tugs are testing with a skimmer system which is part of Clean Seas' arsenal of oil booms, skimming boats, dispersants, foams, and other equipment to contain and combat spills.

granted them a temporary injunction enjoining the Secretary of Interior from issuing construction permits. The suit would delay pipeline construction for five years and escalate the estimated cost from $1 billion to $8 billion.

Environmentalists began putting political roadblocks in the way of all action and planning for energy growth. Their targets were not only the Alaska pipeline and offshore oil, but also the construction of oil refineries and nuclear power plants, development of oil shale and strip coal mines.

Congress also passed the Clean Air and Clean Water Acts of 1970, setting stringent new antipollution standards, causing power plants to convert from high-sulfur coal to oil. Oil imports rose

When the federal government announced in the summer of 1970 it had alerted its Emergency Petroleum Supply Committee to be ready for a possible world oil emergency, newspapers carried front-page stories about a winter fuel crisis, shortages, higher fuel prices, industrial shutdowns, and worker layoffs. Government, industry, and the public were caught by surprise. Suddenly, national and international energy developments had collided to bring about a situation like one of those disasters occurring on a turnpike in foggy weather when one car runs into another, and soon there is a massive pileup.

In May 1970, a bulldozer in Syria "accidentally" cut Tapline, a pipeline carrying half a million barrels of oil daily from Saudi Arabian fields to the Mediterranean, crossing through Jordan, Syria, and Lebanon. Syria refused to let repairs be made until higher transit fees were paid. The pipeline would remain closed for nine months. Oil had to be shipped the long way around Africa to European and American markets, creating a world tanker shortage.

A month after Tapline closed, Libya, whose constitutional monarchy had been overthrown in September, 1969, in a military coup led by Col. Muammar Qadaffi, cut back American oil companies' production by 800,000 barrels daily in a power play to force them to raise their prices and taxes paid to Libya. This critically reduced fuel oil imports to the industrialized U.S. East Coast which was dependent on them for 94 percent of its needs. American industry and government emergency measures managed to avert the predicted fuel oil emergency, but an international energy Pandora's box had been opened that could never be closed again.

In July 1970, the National Petroleum Council, official industry advisory committee to the Interior Department, published a two-year study, headed by Ira H. Cram, one of the world's leading petroleum geologists. Cram coordinated research by the nation's most experienced industry and government geologists on America's oil and gas potential. They estimated future oil production could be four times that of the past 111 years and natural gas pro-

EXXON

Exxon's 200-million barrel Hondo field was discovered in July 1969 in the Santa Barbara channel. The production platform was expected to be in operation and producing 21,000 barrels daily within five years. However, due to federal and state Environmental Impact Studies, public hearings, and lawsuits resulting from the 1970 National Environmental Protection Act, it took twelve years before the first barrel could be produced in 1981.

The wildcat being drilled in the Libyan desert, shown above, was one of the many which catapulted Libya, from a nation with no known resources in 1959 as its lands were 90 percent barren desert, into the world's third largest oil-exporting country by 1969, exceeding Saudi Arabia's exports by 100,000 barrels a day. Oil was first discovered by Exxon and developed by twenty companies, mostly American. Although Americans were more interested in landing on the moon in 1969, events in Libya would help precipitate the energy crises of the United States during the 1970s.

duction could be 3.6 times past production if government policies encouraged exploration. However, the 1970 National Environmental Protection Act had already precluded rapid development of such abundant domestic resources. Furthermore, nationwide price controls imposed in 1971 were removed in 1972 on everything *except* on all oil production as of 1971, further drying up exploration capital. Between 1970 and 1973 gasoline demand spurted ahead of supply at a record rate. Imports soared and supplied 37 percent of consumption by 1973.

Fuel shortages during the winter of 1972–73 were severe enough to hit every part of the country. Workers were laid off when plants and factories could not get enough gas or heating oil. Schools and universities closed from New England to Colorado to Texas. Grain rotted in western silos for lack of heat. Major airlines rationed jet fuel. When Congress convened in January 1973, more than a thousand energy-related bills were introduced. However, next to energy, the nation's biggest shortage was a shortage of action. Congress busied itself trying to fix the blame rather than to prevent shortages. The oil industry was severely attacked on charges that major companies had conspired to raise prices and put independents out of business. From January to September, the only government energy action was abolishment of import quotas by President Nixon and an order to increase the amount of federal offshore acreage offered for oil exploration. Nixon urged Congress to pass legislation to provide incentives to increase domestic energy supplies.

Meanwhile, the Libyan cutback in 1970, which helped precipitate the first major U.S. potential energy crisis, sparked a Middle East oil revolution. For ten years a cartel of Arab oil countries, Iran, and Venezuela, known as the Organization of Petroleum Exporting Countries, or OPEC, had been struggling to wrest control of their oil resources from international oil companies which had developed concessions on a fifty-fifty profit-sharing basis. Libya succeeded in increasing profits to 55 percent. Then, from 1970 to September 1973, the OPEC countries successfully began dismantling the international oil industry with a series of nationalizations, takeovers, forced government participation in oil company operations, increased taxes, and a 12 percent boost in oil prices.

In September President Nixon threatened Congress that if it did not act on his energy proposals America "would be at the mercy of the producers of oil in the Middle East." It was too late.

AMERICAN PETROLEUM INSTITUTE

Above, one of the villains in the energy crisis scenario.

SOCAL

The California freeway, left, symbolized what happened to gasoline demand. In 1968, the number of passenger cars, trucks, and buses in America passed the 100 million mark and traveled 1 trillion miles. In 1973, 96 million passenger cars alone traveled 1 trillion miles, and an additional 275 billion miles were driven by more than 22 million trucks.

AMERICAN PETROLEUM INSTITUTE

AMERICAN PETROLEUM INSTITUTE

SOCAL

Americans observed government-urged conservation measures—gasless Sundays, lowered thermostats, reduced speed limits—but when shortages became real, prices increased, and gasoline lines formed nationwide, such as the one at right, the national mood turned ugly.

Arab countries, frustrated by the inaction of the United States to influence Israel to withdraw from territories occupied in the 1967 war and make a peace treaty, decided on the shock treatment of war. Egypt and Syria launched a massive surprise invasion of Israel in October 1973. After two weeks of bloody fighting, the Arabs dropped a political and economic bomb whose spectacular fallout spread around the world. They cut back one-fourth of their oil production and totally embargoed shipments to the United States, the Netherlands, Portugal, South Africa, and Rhodesia because of their pro-Israel policies. Only 5 percent of American consumption was supplied by Arab oil. However, Arab oil supplied Western Europe with 73 percent of its needs and Japan with 45 percent. The Arabs hoped Europe and Japan would pressure America to intervene politically. The United Nations arranged a cease-fire in three weeks, but the international energy war intensified.

Oil imports into the United States did not dip because of the Arab embargo and cutback until December. However, the huge OPEC price increases quickly hit the American consumer since one out of every three barrels of oil consumed was imported, and price control regulations permitted increased costs of raw materials to be passed along to consumers.

The Arab oil embargo ended in March 1974, but OPEC's quadrupling of foreign oil prices had skyrocketed gasoline prices by ten to fifteen cents a gallon in a few months. When major oil companies announced 1974 first-quarter profits were 78 percent higher than the same 1973 period, the public was convinced the crisis resulted from "oil profiteering." Almost nobody believed the companies' explanation that 75 percent of increased revenues occurred outside the United States and that 85 percent of profits growth came from abroad. Nor did the public believe that oil companies only made an average profit of two cents a gallon on gasoline made from the mix of domestic and foreign oil and that their return on investment averaged the same as all U.S. manufacturing companies. Also, major companies were investing twice as much in the United States as they were abroad. Their capital expenditures in the United States for 1973 and 1974 were two and one-quarter times as large as their profits.

Although Congress did nothing in 1974 but hold hearings, positive actions to increase domestic supply and to find new resources were under way. When gasoline lines formed, the Cost of Living Council authorized a dollar increase in the controlled price of oil to provide exploration incentives. The price of uncontrolled new oil increased by four

AMERICAN PETROLEUM INSTITUTE

Three days after the October 1973, Arab-Israeli War began, the Arabian Gulf Arab country members of OPEC, shown at left, unilaterally increased oil prices a staggering 70 percent, or about two dollars a barrel. Libya leapfrogged, boosting prices by 93.8 percent. Then, in December, the Gulf Arab states and Iran jumped the price of oil by 130 percent over the 70 percent jump of two months before. This meant that the oil-consuming nations' import bill in 1974 would rise by an astronomical $95 billion over 1973.

STANDARD OIL (INDIANA)

When Congress convened in January 1974, it was in a punitive mood. A blizzard of 3,000 energy bills were introduced in the Senate and House and hearings began, such as the one above, of the Senate Energy Committee, to determine who and what was responsible for the energy crisis.

dollars a barrel. Independents began plowing back profits in the most vigorous oil and gas exploration program the country had seen in fifteen years.

Most importantly, in November 1973, immediately after the embargo, Congress had ended the five-year environmental impact court battle over the Alaskan pipeline by authorizing its construction.

In environmentalists' opposition, caribou became the symbolic victim of the pipeline to Americans who had never visited Alaska. However, caribou, with herds totaling 440,000, are slightly more than Alaska's people population. Their migratory patterns are such that only 6 percent of the caribou normally cross the pipeline right-of-way en route to calving grounds. Caribou can migrate forty miles a day, grazing along the way. On the aboveground portions of the pipeline, Alyeska built ample overpasses, but the caribou seldom bothered to use them. They just go under the six- to eight-foot elevated pipeline. They pay little attention to oil development or the pipeline. On one summer day some 6,000 caribou passed through the Arco oil field and grazed around the rigs for several days. They had to be herded off airplane runways so planes could land. The Central Arctic caribou herd, which summers around Prudhoe Bay, actually increased its numbers from 5,000 in 1969 to 6,700 in 1979.

Environmentalists were also extremely concerned about the danger to certain rare species in the Arctic, particularly peregrine falcons and Dall sheep. Peregrine falcons, nesting in regions near the pipeline, were far enough away not to be disturbed. Although Dall sheep lamb and graze in one pipeline area, construction was timed to minimize disturbance to them.

In developing Prudhoe Bay and building the pipeline, the oil companies made one of the most in-depth environmental studies ever made on any one area in the world. Then, in carrying out their operations, they demonstrated how compatible the protection of the environment and energy development can be.

ALYESKA PIPELINE SERVICE COMPANY

The trans-Alaska pipeline is one of man's greatest engineering feats and the most heavily researched, safest, and best-engineered pipeline ever constructed. The 800-mile line from Alaska's top to Valdez, the ice-free port of its southern coast, occupies only 8.2 square miles out of Alaska's 586,412 square miles. It is comparable to building a highway to transport oil. In order to avoid thawing the ground, half of the pipeline is elevated above ground in areas where the soil contains too much ice, as shown above near Fairbanks. The zigzag configuration permits the forty-eight-inch-diameter pipe to expand and contract without breaking or buckling in broad variations of temperature. Once Congress gave the green light, construction began in mid 1974, and it was finished in a short three years.

ALYESKA PIPELINE SERVICE COMPANY

More than 72,000 men and women contributed their strength and skill to the rigorous job of building the pipeline.

AKYESKA PIPELINE SERVICE COMPANY

Permafrost posed one of the major engineering problems. Almost 85 percent of Alaska is underlain by permafrost, which is permanently frozen material, anything from solid rock to muddy ice, extending from a few feet to hundreds of feet below the surface. In the coldest regions, permafrost is close to the surface, topped by a layer of soil and vegetation which is subject to seasonal thawing and freezing. This is called tundra. In winter, when it is frozen, it can be traversed easily. However, when traversed during the short summer thaw, the ruts become permanent when the tundra freezes again. Consequently, thick gravel pads must be built to insulate the permafrost from thawing in order to construct roads, drilling sites, and buildings. At left, a geologist examines exposed permafrost along the pipeline route.

A vital part of construction was building a 360-mile road from Prudhoe Bay through the wilderness to the Yukon River to transport equipment and supplies. At right is the final linkup of construction crews on the Koyukuk River, about 100 miles north of the Yukon. A total of 32 million cubic yards of gravel was used to protect the permafrost.

ALYESKA PIPELINE SERVICE COMPANY

A major problem in aboveground construction was how to keep the steel, vertical pipeline supports from either settling or raising in the ground through thawing or frost heaving. The high Arctic was no problem as the ground remains frozen tightly in place even during the short summer. Further south, the unique solution was to keep the ground frozen by inserting two hollow tubes in each support, containing a few pints of liquid ammonia and ammonia vapor. When the temperature is low, the vapor condenses into liquid ammonia and runs down to the bottom of the tube. There it absorbs heat, turns to vapor again, rises to the top, and endlessly repeats the cycle, each time carrying heat from the warm ground. The dual heat pipes protruding from the supports, shown below, remove so much heat in winter that the ground remains frozen all summer.

ALYESKA PIPELINE SERVICE COMPANY

ALYESKA PIPELINE SERVICE COMPANY

Stretching across rolling tundra and rugged mountain country, the pipeline passes over and under more than 800 streams and rivers. Above is the Tazlina River crossing, one of the thirteen major river bridge crossings. Neither construction nor subsequent operations caused any damage to the salmon and other fish populations.

During construction, all disturbed areas along the 800-mile pipeline corridor, except roadbeds, were the target of a massive revegetative effort as illustrated by aerial reseeding, shown below. A principal objective was to control soil erosion and protect the pipeline. Extensive research was done to develop the best seed mixtures.

ALYESKA PIPELINE SERVICE COMPANY

AMERICAN PETROLEUM INSTITUTE

The potential fate of Alaska's caribou herds migrating across the pipeline right-of-way was a key factor in environmentalists' opposition to building the Alaska pipeline. However, the caribou paid little attention to the pipeline. The Central Arctic herd summering around Prudhoe Bay increased from 5,000 in 1969 to 6,700 in 1979.

SOHIO

The sixteen owners of Prudhoe Bay leases unitized operations to avoid duplication of facilities and to ensure maximum efficient production with maximum environmental protection. Sohio Alaska Petroleum Company, owning almost 53 percent of the field, operates the western part and Arco, with 20 percent ownership, operates the eastern part. At left, is the Sohio operating area. All buildings are on top of pilings, leaving at least a three-foot gap between the building floor and gravel surface beneath it, further insulating permafrost from building heat. Sohio's Prudhoe Bay reserves, combined with its "lower 48" states reserves, give it more proven reserves in the United States than any other company. What is essentially an offshore drilling technique is used on the North Slope. Wells drilled from gravel pads each tap 160 acres. Drilled vertically through 2,000 feet of permafrost, they angle out to each well's target area. By 1982, Sohio drilled 221 wells and Arco drilled 215; with production peaking at 1.5 million barrels daily. The field's vast natural gas reserves were being reinjected into the ground pending the economic feasibility of building a multibillion dollar, proposed natural gas pipeline to the lower forty-eight states.

No matter how severe the weather is outside, Prudhoe Bay workers live in a country club atmosphere, as demonstrated by the space-age design of the three-story high, skylighted ceiling of the Sohio's Operations Camp dining room at right, where four gourmet meals are served every twenty-four hours. The 570 workers, who spend eight out of every fourteen days at Prudhoe Bay working twelve hour shifts, also enjoy an arboretum with trees and flowers, saunas, game rooms, gymnasiums, swimming pool, movie and taped television theater, library and attractive bedrooms.

SOHIO

ATLANTIC RICHFIELD

Valdez, the beautiful, northernmost ice-free harbor in the United States, is the 1,000-acre marine terminal of the Alaska pipeline, shown above. Its eighteen storage tanks have a total capacity of 9,180,000 barrels of oil. It receives an average of fifty-three tankers a month to deliver oil to the West Coast and to the U. S. Gulf of Mexico coast and eastern ports via the Panama Canal. The terminal cost $1.4 billion, bringing the total cost of the pipeline and terminal to $9.4 billion, the largest privately financed construction project ever undertaken. Overall investment by the oil companies, with Prudhoe Bay leases, was approximately $15 billion, including all field development and pipeline costs, interests on the pipeline investment, and the cost of tankers carrying the oil. Ironically, if the pipeline had been completed in 1972, as first planned before environmentalists stopped its construction, the 1973 Arab oil embargo would not have happened as Alaskan oil would have replaced Arab oil imports. The first shipment of Alaskan oil was August 1, 1977.

PHILLIPS PETROLEUM

Federal legislation permitting construction of the Alaskan pipeline requires congressional approval for exporting Alaskan oil to countries not adjacent to the United States, approval that has not yet been given. However, Alaska's first oil and gas fields were found in southern Alaska on the Kenai Peninsula in Cook Inlet between 1957 and 1962. Phillips Petroleum built America's only liquified natural gas plant in Kenai, shown at right, in 1969 which ships LNG to Japan.

Energy problems took a back seat the last half of 1974. The nation was in turmoil over President Nixon's August resignation, President Gerald R. Ford's September pardon of Nixon, and the Watergate trials of former Nixon aides.

In January 1975, President Ford called for lifting price controls from domestic oil, imposing a "windfall" profits tax on producers and import fees on foreign oil. However, the Democratic Congress passed a new Tax Reduction Act in March which wiped out the controversial percentage depletion tax allowance—originally designed to provide exploration incentives—for all but the smallest operators. The bill removed approximately $2.5 billion annually from oil industry working capital. There was an immediate cutback in planned oil and gas exploration. Then, in December, Congress passed, and President Ford reluctantly signed, a new Energy Policy and Conservation Act which not only imposed another forty months of price controls on old oil, but also extended them to new oil and rolled back free market prices of $12 to $13.50 a barrel to $11.28. The act reduced industry's income by another $3 billion annually, with almost half of that coming from pockets of independent producers.

In January 1977, President Jimmy Carter was inaugurated in the midst of the century's worst winter, with weather and natural gas shortages throwing more than 1.8 million people out of work. Carter promised an energy program which would be "the moral equivalent of war." However, his plan was, in essence, a massive, complicated taxation program to force conservation by a federal tax on production to increase consumer prices, but to continue price controls on producers. There were almost no incentives to increase energy supplies. By this time, the United States was importing almost half of its oil needs and running a huge national trade deficit. It imported $45 billion worth of oil in 1977 compared to $7.7 billion in 1973. The most alarming fact was that public opinion polls showed that half of all Americans didn't know the nation imported *any* oil and did not believe there was an "energy crisis."

Congress and President Carter fought bitterly over his various energy proposals during his four-year term. Congress gave him only part of what he

U. S. CAPITOL HISTORICAL SOCIETY

From 1975 to 1980 Congress, shown in session above, did more to discourage development of energy supplies than to increase them, and the nation went from one energy crisis to another.

wanted and a lot he didn't. In 1978, Congress passed an act phasing out price controls on natural gas by 1985 with sizable price increases in the interim. New gas below 15,000 feet was exempt from price controls. However, it put intrastate gas under price controls for the first time.

The public was stunned by a new crisis in 1979 when gasoline lines began to form nationwide and prices jumped 50 percent to about one dollar a gallon. The reasons were international. Iranian oil production had been shut down in late 1978 and early 1979 in the struggle between the Shah of Iran and Ayatollah Khomeini. There was a world oil shortage: U.S. imports dropped, and OPEC seized

The 1979 gasoline lines, shown above, were like the rerun of a 1973 bad old movie.

President Jimmy Carter signs the 1980 "windfall profits" excise tax bill on oil production.

the opportunity to double world oil prices between October 1978 and June 1979. By late summer, gasoline supplies normalized, but the penalty of importing half of U.S. oil needs kept prices high.

Oil price controls were due to expire September 1981, but Carter exercised his authority to phase out controls over twenty-eight months beginning in June 1979. He did this not because he believed it would increase domestic oil and gas supplies, but to pressure Congress to pass a "windfall profits" excise tax. Congress had refused to give him his original demand for a crude oil wellhead tax on producers, but he still wanted higher U.S. prices as a means to obtain conservation, with government getting the bulk of the revenues. Carter lashed out at the oil industry, accusing it of "profiteering." However, Congress continued to balk. Finally, in 1980 it gave him a "windfall profits" tax bill which would yield the government $277.3 billion in revenues from 1980 through 1990, with major companies paying $204.8 billion and independents, $22.5 billion.

Between 1979 and 1982, there were dramatic changes in U. S. oil and gas exploration and consumption due to higher prices and phasing out of oil and gas controls.

Despite the huge tax bites of elimination of percentage depletion tax allowances and the "windfall profits" tax, there had been a steady improvement in oil and gas prices received by producers. Major companies and independents ploughed their profits back into the ground in the sharpest increase in exploration and development drilling the industry had ever known. In 1978, the industry drilled 48,513 wells, but in 1980 it drilled a record-breaking 62,462. In January 1981, one of President Ronald Reagan's first acts was to eliminate all oil price controls instead of waiting for them to expire in September. Industry response was immediate. It broke its own record by drilling 77,500 wells in 1981. The industry spent a record $74.7 billion in 1981, of which exploration and production accounted for two-thirds. This capital outlay was more than double that of 1978. The steady decline in domestic production since 1974 was halted and an upward trend began.

With decontrol, domestic crude oil prices leaped to match OPEC world prices. This had a pro-

PHILLIPS PETROLEUM

AMERICAN PETROLEUM INSTITUTE

Gasoline production accounts for half of crude oil consumption. Higher prices induced Americans to conserve gasoline by cutting down on unnecessary travel and van pooling to work, as illustrated above left. Detroit began production of compact cars getting more miles to the gallon. Twenty percent of total energy demand is consumed in homes, and Americans learned they could save as much as 10 percent on heating bills by turning down thermostats, as shown above right. In summer, each degree the thermostat is turned up saves about 5 percent of electricity energy used for cooling.

The biggest slice of the total energy pie—40 percent—is consumed by industry. By good energy housekeeping, industries found they could cut down consumption by 15 percent. So did homeowners, as illustrated below with insulation of storm windows, left, and building insulation, right.

AMERICAN PETROLEUM INSTITUTE

AMERICAN PETROLEUM INSTITUTE

AMERICAN PETROLEUM INSTITUTE

found impact on imports. Conservation became a reality when oil product prices increased. In 1978, imports of 8.2 million barrels daily accounted for almost half of U.S. consumption. Demand began to drop sharply in 1980, and in 1982 imports had decreased 40 percent to 4.9 million barrels daily. Import costs which had been $73.2 billion in 1980 were down to $55.2 billion.

Although the nation was still vulnerable to any major disruption of import supplies, particularly from the unstable Middle East, the surge of exploration drilling had proved that price controls were the enemy of increased domestic supplies of oil and gas and that the American energy problem was basically a political one. Natural gas had not yet been completely deregulated, but accelerated development of the nation's abundant oil and gas potential had begun after being derailed for so many years.

During the 1974 to 1979 crises, there was great publicity about developing energy sources other than oil and gas. The Carter administration pushed for a crash, multibillion dollar government-financed program to develop synthetic fuels from the nation's vast coal and shale oil resources. Technology was in

PHILLIPS PETROLEUM

A great part of oil industry research is concerned with new ways to conserve oil. Phillips Petroleum invented a new asphalt-paving process, shown above, which cuts air pollution and uses 30 percent less petroleum. The old method liquefied asphalt for handling by using kerosene and other petroleum solvents. When hot asphalt was applied to the roadbed, the solvents escaped into the air causing pollution and representing a loss of as much as 336 million gallons of oil a year. Oil doesn't normally mix with water, but the new process emulsifies asphalt with water. When laid down, the water evaporates, and a solid highway surface is formed.

TEXACO INC.

After four decades of research, Texaco has developed a new type of vehicle engine based on the Texaco Controlled Combustion System which gives a fuel economy of more than 35 percent compared to the conventional gasoline engine. Furthermore, it can use a wide variety of fuels or a wide-boiling-range fuel which uses less crude oil to make than does gasoline or diesel fuel. Conventional engines can be adapted to use the new system. United Parcel Service, which operates the world's largest fleet of commercially owned vehicles, is testing the new engine, shown left, with plans to convert about three-fourths of its 50,000 trucks.

"The rock that burns" is what American Indians called shale oil, as demonstrated at left. The enormous shale oil reserves in the western United States, 80 percent of which are government owned, insure the nation's energy future. Technology is in place to produce them, but they are not yet economically competitive.

TEXACO INC.

place to begin using western states' oil shales, which contain an estimated five times the world's producible oil reserves. However, as the 1980s began, a surplus of oil and gas and decreased demand made it uneconomic to produce high-priced synthetic fuels. Most projects were abandoned or postponed, with the exception of Union's $500 million plant in Colorado, scheduled to begin producing 10,000 barrels of oil daily in 1983. New techniques for producing synthetic fuels from coal were being developed; however, coal mining was hampered by environmental restrictions.

The expansion of nuclear energy, which supplied only 3 percent of U.S. energy in 1980, had been brought to a halt following the Three Mile Island accident in 1979. Development of solar power technology was in its infancy except for minor uses.

In the 1980s, oil and gas still remained the name of the energy game in the United States and worldwide.

While the United States went through its energy agonies of the 1970s and nondevelopment of oil and gas resources, American major and independent oil companies were discovering major new resources worldwide. The biggest oil exploration surprise occurred in 1970, when Phillips Petroleum Company announced a major oil strike in Norway's offshore waters of the North Sea. A huge new oil province had been discovered in a 600- by 350-mile sea whose waves lapped the shores of some of the world's most oil-hungry nations—Great Britain, the Netherlands, Germany, Denmark, and Norway. It became the richest, busiest, most accessible target for big oil production since the first Middle East discoveries, with U.S. companies dominating its development. One giant oil field after another has been

TEXACO INC.

Coal is America's most abundant energy resource. America has an estimated 3.2 trillion tons, one-fifth of estimated total world coal resources. However, during the 1970s it supplied only an average of 18 percent of the nation's energy. Its development and use is hampered by federal and state air standard and strip-mining regulations. However, new techniques are being developed to convert coal into a wide range of clean, liquid fuels and synthetic gas. At left is the nation's first integrated coal-gasification-combined-cycle-generating plant in an existing utility system in California. Texaco developed the coal gasification process, and electricity will be generated at southern California Edison's generating station, using General Electric's combined-cycle technology. The $300 million plant begins operations in 1983.

The oil industry is pioneering the development of the nation's geothermal resources by drilling wells to produce wet steam and hot water, created in the earth's crust by its natural heat, to use in generating electricity. The first commercial plant was built in California in 1960. Development began accelerating in the 1980s, not only in California, but in Utah, Nevada, and New Mexico. The U. S. geothermal resources are estimated to have the equivalent of 700,000 barrels daily of crude oil, or about 8.5 percent of 1981 U. S. oil output. In California, geothermal energy has the capability of meeting 25 percent of the state's electrical demand if developed. At right, a geothermal well is tested in Phillips Petroleum's geothermal field in Utah, which begins production in 1984.

PHILLIPS PETROLEUM

Above is the heart of the Greater Ekofisk Development in the Norwegian North Sea where Phillips Petroleum found the North Sea's first oil in 1970. This $6 billion complex is operated by Phillips, with nine partner companies, to handle production from seven fields.

PHILLIPS PETROLEUM

found. Probable reserves are estimated at 40 billion barrels.

The North Sea exemplifies the unpredictability of oil and gas exploration. Wells were drilled around North Sea land fringes for forty years, discovering uneconomical accumulations of oil or gas. Not until 1959, when Holland's supergiant Groningen gas field was discovered did geologists begin to suspect offshore might be elephant-hunting country. North Sea drilling has been the industry's greatest technological challenge as it is the world's harshest offshore environment, with men and equipment almost continually buffeted by 100-mile-an-hour winds and pounded with waves up to ninety-five feet. The North Sea was conquered by the extraordinary development of American offshore drilling and production technology.

Finding oil in South America's huge upper Amazon Valley jungles and pipelining it over the Andes Mountains ranks with Arctic oil pioneering as one of man's greatest technological achievements. In the late 1960s, Texaco and Gulf spent $16.5 million wildcatting in Colombia's Amazon area before finding oil on their fifth test. Men, equipment, and supplies to develop the field were air-lifted in a massive helicopter operation. Williams Brothers, the world's leading pipeline builders, built a monumental, 100,000-barrel-daily-capacity pipeline over the 11,000-foot Andean peaks, part of whose construction is shown at left. Oil reached the Pacific in 1969. Then Texaco and Gulf found oil in Ecuador's Amazon Valley, and Williams Brothers tamed the Andes again, laying another equally difficult 250,000-barrel-a-day pipeline.

TEXACO INC.

TEXACO INC.

During the 1970s, development of Indonesia's offshore waters by American independent and major companies made Indonesia the world's eighth largest oil-exporting country and sparked a boom throughout Southeast Asia. Shortly after World War II, Caltex, owned jointly by Socal and Texaco, had developed a billion barrel field, Minas, whose first well is shown at right. Exploration remained in a deep freeze for a quarter-century under President Sukarno's nationalistic policies. When his government was overthrown in 1965, President Suharto put out the welcome mat for foreign investment. General Ibnu Sutowo, head of Pertamina, the state oil enterprise, innovated production-sharing contracts. This concept, providing profit incentives, has been followed by other nations around the world.

Oil was found offshore Dubai, on the Arabian Gulf, by Conoco in 1969. Khazzan I, the world's first ocean floor storage tank is shown at left being towed fifty-eight miles into the Gulf where it was slowly submerged until the bottom rested on the seabed 150 feet below the surface, and then anchored. The tank, shaped like an inverted champagne glass, and as tall as a twenty-story building, is bottomless with a deck and platform projecting 40 feet above the surface. It holds 500,000 barrels of oil and utilizes the principle that oil floats on water. As oil is pumped in, water is forced out of the open base. The reverse process takes place when oil is pumped out into the holds of tankers.

CONOCO

STANDARD OIL (INDIANA)

Although Africa's first oil was found in Egypt in 1911, it was not until Amoco discovered the 2-billion-barrel El Morgan field in the Gulf of Suez in 1965 that Egypt joined the ranks of the potentially oil- and gas-rich nations. Further exploration languished until 1973 because of lack of profit incentives under government policies. However, when Egypt started an open door policy with favorable production-sharing contracts, forty-nine companies from eighteen countries rushed to sign them and to explore onshore and offshore. By 1982 Egypt was approaching the 1-million-barrel-a-day production mark, and oil exports were the nation's chief revenue source. Israel's 1979 return of the Sinai opened up a potentially productive new exploration area. At right is an Amoco wildcat in Egypt's western desert.

The 1980 convoy of "vibroseis" trucks, above, in Utah, symbolizes the amazing advances in geophysical technology, enabling the oil and gas industry to locate complicated geological traps for the discovery of giant oil and gas fields.

NEW OIL AND GAS HORIZONS:
1980s and Beyond

As the 1980s began with a U.S. oil and gas exploration boom following removal of oil price controls and phasing out of natural gas price controls, explorers were eager to develop the nation's abundant oil and gas resources. Claims that America was running out of oil and gas were far from the truth.

The U.S. Geological Survey, which traditionally makes the most conservative estimates, concludes that of total potential oil resources, *recoverable with present technology*, the United States has exhausted 39 percent, has a reserve of 24 percent, and has 37 percent—or another 100 billion barrels—left to discover if industry is given sufficient incentives for exploration and has access to the potential land and offshore areas. Furthermore, the survey points out that another 150 to 235 billion barrels of oil known to be in place can be recovered by the use of expensive new technology called enhanced recovery. The Potential Gas Committee, a government-industry group sponsored by the Colorado School of Mines, estimates that 4.5 times proved reserves of natural gas remain to be found.

Accelerated geophysical and geological research in the 1970s had provided the industry with remarkable new tools with which to find new oil and gas provinces. The application of computer technology to geophysical work and development of lightweight seismic recording equipment were helping geophysicists produce pictures—charts and maps of underground formations—that were barely dreamed of by early petroleum geologists.

Although the principle of seismograph remains the same—recording artificially created "earthquake" soundwaves from rock formations below to determine their configuration—the means of creating sound waves and methods of recording results are drastically different from seismograph invented

STANDARD OIL (INDIANA)

Rapos, or Rapid Propane Oxygen System, was invented by Standard Oil of Indiana to map subsurface rock formations at less cost in difficult terrain or where environmental considerations restrict use of other seismic systems such as dynamite. The one-man operated unit, shown right, is hinged, allowing it to "bend" around trees and other obstructions. It weighs only as much as an average farm tractor. Instead of using dynamite, it creates sound waves to bounce off the rocks and be magnetically recorded, by boring an auger ten feet in the soil and detonating a propane-oxygen explosive.

TEXACO INC.

Geophysical data collected by crews on land and offshore is recorded on magnetic tape reels which computers transform into many cross sections and contour maps. Racks of magnetic tape reels line the walls of the computer room, above, in Texaco's Geophysical Data Processing Center at Bellaire, Texas. Twenty-four hours a day, the system is fed enormous amounts of geophysical information for processing from Texaco's exploration operations throughout the world.

TEXACO INC.

Texaco data-processing analysts, left, study a seismic section, produced by the computer to identify such subterranean features as faults or dome-shaped anticlines that may trap oil. Exploration requires the coordinated efforts of geophysicists, geologists, paleontologists, geochemists, well-log analysts, computer scientists, mathematicians, and many others to assemble all the clues to decide where to drill a wildcat well. However, even with today's sophisticated technology only the drill bit can determine if oil or gas is there. Only 1 wildcat well out of 9 finds an oil field. In the United States, only 1 in 68 wells results in a significant discovery—more than 1 million barrels of oil or 6 billion cubic feet of natural gas.

in the 1930s. In 1956, Conoco invented "vibroseis," which, unlike the use of dynamite as an energy source, merely shakes the ground. It developed slowly, but is now so effective and has such little impact on the environment that by 1984 vibroseis crews are expected to exceed dynamite crews by 50 percent. The method involves large trucks in tandem. Stopping in formation, each truck lowers a giant metal plate from its underbelly, lifting the several ton vehicle off the ground. Then, synchronized by computers, hydraulic mechanisms vibrate the metal plates in unison, sending shock waves into the earth. As the signals bounce back, they are picked up by geophones, laid along the shot line, and transmitted to a computerized recording truck a short distance away. The vibrating trucks slowly lower to the ground, advance in formation another few feet, and repeat the sequence until the length of the survey line is covered. The geophone readings on magnetic tape are processed by the computer to display the shock waves in seismic sections to be analyzed and converted into maps in a processing center.

The newest technique is three-dimensional seismology. Portable, lightweight cassette recorders use radio telemetry instead of electrical cable to relay seismic reflections picked up by ultrasensitive geophones. Three-dimensional pictures of the rock formations are achieved by vastly increasing the number of subsurface samples taken and data obtained so that the computer can convert them into graphic displays. This involves working with over 1,000 channels of information instead of the 96 used in conventional seismology.

Running from Alaska to Central America, the western Overthrust Belt is believed to have resulted from collision of the Pacific and North American tectonic plates more than 130 million years ago. Sheets of sedimentary rock as large as twenty miles wide and four miles thick were thrust up and over or down and under other strata. The violent geological occurrence created the majestic Rocky Mountain ranges and foothills. Underground it created bizarre structural features which, until recently, defied any kind of geological mapping.

Oil seeps in Wyoming indicated there was oil, but about 300 exploration wells drilled over the years were dry holes until, in 1975, Quasar Petroleum Company, a Texas independent, discovered a giant oil field in Utah, setting off the western Overthrust Belt exploration boom. By the end of 1981, eighteen oil and gas fields had been established along the Wyoming-Utah-Montana part of the thrust belt. At least six of them rank as giants with reserves of more than 100 million barrels each or 1 trillion cubic feet of natural gas per field.

Geologists believe that production is possible all along the U.S. Overthrust Belt. Important oil and

STANDARD OIL (INDIANA)

Above, a helicopter-supported three-dimensional seismic survey is conducted in the Whitney Canyon area of the Overthrust Belt, in Wyoming.

MARATHON

By 1981, the oil and gas potential of the Overthrust Belt of the western U.S. was being compared in importance to the discovery of the Prudhoe Bay field in Alaska. Opening up this huge new oil and gas province has been made possible primarily by the remarkable new seismic techniques as its complicated geology makes it one of the most challenging exploration areas in the world. Above, geologists examine rocks in the Utah portion of the western Overthrust Belt.

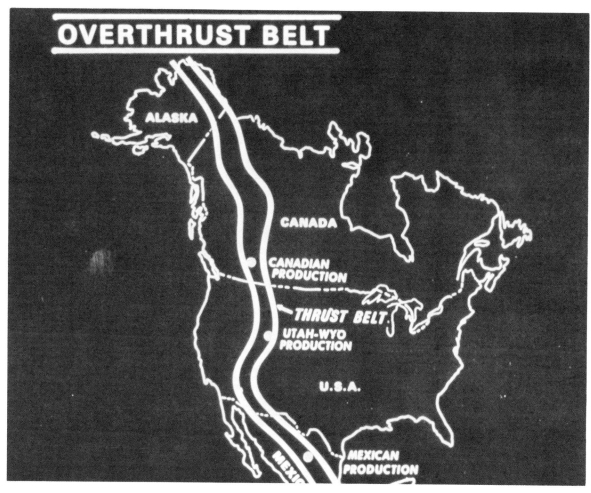

AMOCO

gas discoveries have been made in thrusted areas of Canada and Mexico.

In 1981, the western Overthrust Belt was still 95 percent unexplored. A major reason is that the principal landlord in the region, the U.S. government, has not been granting drilling permits in wilderness areas or areas that may be designated wilderness in the future. The government owns 60 to 70 percent of the Overthrust Belt, including 73 percent in Arizona and Utah, 65 percent in Idaho, and 62 percent in Wyoming.

As recently as 1968, most federal lands were open to economic development, but the government has staged a land rush in reverse. By 1977, three-fourths of the federal domain—or an area nearly equal in size to all the states east of the Mississippi—either had been withdrawn or so burdened by restrictions as to preclude energy-minerals exploration and development. From 1977 to 1979, the Carter administration locked up 56 million more acres of land in the lower forty-eight states and 110 million acres in Alaska.

Ninety percent of the government's 738 million acres, most of it lying in eleven western states and Alaska, are restricted in some degree from oil and gas exploration. The Reagan administration favored opening up wilderness areas for development, which is permitted under law until the end of 1984. But in the face of violent opposition from environmentalists, in 1982 it placed a moratorium on any leasing until the end of 1983. It also proposed legislation to ban leasing on 24 million acres of designated wilderness lands in the lower forty-eight states until the year 2000. At the same time, another 40 million acres of wilderness study areas would be subject to quick review and mineral evaluation and could

Exploration in the Utah-Wyoming portion of the western Overthrust Belt is difficult. Many sites are on or near high ridges and mountaintops. In many areas there are no roads, so routes have to be staked and temporary access roads built. Drilling crews and supplies to support them must be transported long distances from towns and staging locations. The formidable winters of the high country present additional obstacles. The Marathon exploratory well in Utah, at right, is being drilled in typical rugged terrain. No roads back to this remote valley existed before drilling began.

MARATHON

CONOCO

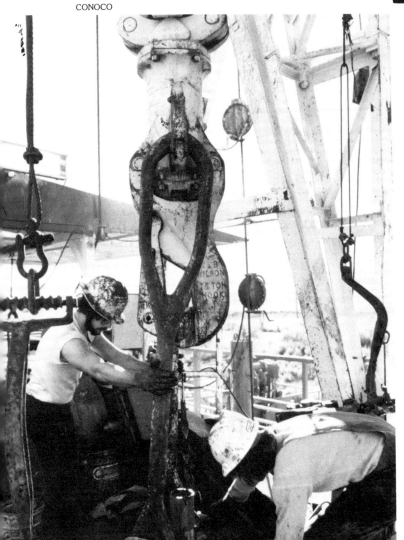

Exploration and development wells are expensive in the Overthrust Belt. Wells typically cost more than $3 million. Drilling is difficult because of hard rocks and steeply dipping formations. A 21,000 foot hole costs from $8 million to $15 million to drill. At left, roughnecks drill for natural gas on a Conoco well in the Ouray Field in eastern Utah.

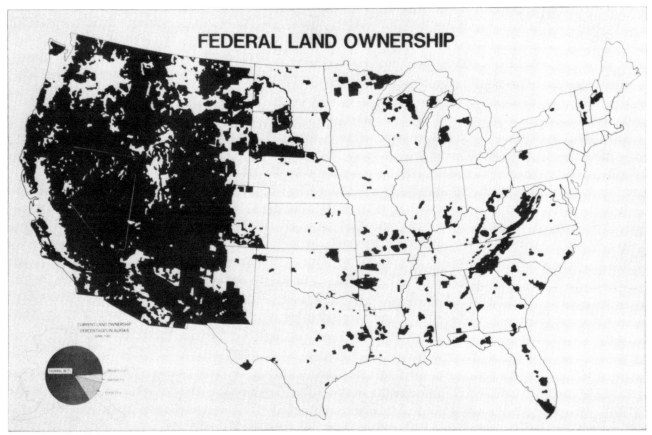

Of the total land area of the United States, 761 million acres, or one-third, is in federal domain. These public lands contain some 50 percent of the nation's low-sulfur coal, 75 percent of the oil shale, 85 percent of the tar sands, 15 percent of developed crude reserves, and a huge potential of undiscovered oil and gas. In addition, the lands contain vast mineral and timber resources.

either be added to existing wilderness acreage or released for mineral development. The highly controversial "war of the wilderness" continued to be as locked up in Congress as the lands themselves.

Oil explorationists maintain that their operations in the Overthrust Belt have demonstrated that energy resources can be developed with little impact on the environment. Small, isolated locations and temporary roads necessary to get to them are restored and do not scar the land. Pipelines are buried. Animal life remains unaffected in the midst of oil exploration and drilling.

A new era in exploration of the huge potential of U.S. offshore oil and gas resources began in the 1980s with the combination of doubling of oil prices in 1979 and the Reagan administration's "open door" policy on federal leasing of the outer continental shelf.

The U.S. Geological Survey conservatively estimated in 1980 that undiscovered recoverable resources in all offshore areas were 38 billion barrels of oil and 139 trillion cubic feet of gas. Despite this great potential, the United States has leased less than 5 percent of its shelf area while other non-Communist countries with coastal areas have leased more than 40 percent of theirs.

Following the Santa Barbara oil spill and the National Environmental Protection Act of 1970, and throughout the decade, federal lease sales were so slow and limited, due to environmentalists' legal actions and political pressures, that, in 1980, the American Association of Petroleum Geologists reported that if sales continued at such a snail's pace it would require more than a thousand years to complete the leasing and exploration of the United States offshore.

After the 1973 Arab oil embargo, the Nixon administration ordered tripling of federal lease sales, but under the Carter administration those scheduled sales were postponed from four months to three years. However, in 1976 the oil industry was given the opportunity to lease some prime frontier areas in the Gulf of Alaska and in the Baltimore Canyon, offshore Atlantic. Exploration in the latter area was delayed eighteen months by environmentalists' legal actions. It took seven years of legal battles and court injunctions before lease sales and exploratory drilling finally got under way in the Georges Bank trough offshore Massachusetts in 1981.

The offshore U.S. West Coast was already known to have the prospect of giant oil fields. Under the Reagan administration, the 1981 sales of leases off northern California broke all state records, including the largest single bonus ever paid for an offshore tract—$336.6 million. Since 1953, when federal outer continental shelf leasing began until 1980, the government received $41.3 billion in bonus bids, royalty, and rental payments from the industry. The values of oil and gas produced came to $62.8 billion, so the oil companies received $21.5 billion, half as much as the government, out of which they had to pay exploration, production, and transportation costs, plus taxes.

In 1982, the Reagan administration announced plans for the most extensive offshore oil- and gas-leasing program in the nation's history, making nearly the entire outer continental shelf—almost 1

Success of the Rocky Mountain thrust belt focused new attention on geologic relationships of mountains and sedimentary basins. The 1,100-mile Eastern Overthrust Belt, along the Appalachian Mountains from New York to Alabama, became a new exploration target with majors and independents leasing along its entire length. Wildcatting began in 1980. Geologists predict that discoveries will be gas prone, based on thrust-related discoveries already made and the geologic age of the rocks which are older than the Rocky Mountains area. The Texaco eastern Overthrust wildcat, shown below, is a 19,500-foot, $14 million test. The crew is examining a $15,000-drill bit, one of sixty-two used. The well is about 100 miles from the 1859 Drake discovery well that started the American oil industry. Colonel Drake's drilling tools cost $76.50 to drill 69½ feet.

TEXACO INC.

The federally owned U.S. outer continental shelf and slope, which begins three miles from the coastline and contains more than 1 billion acres, is one of the nation's major potential oil and gas resources.

billion acres—available for leasing over the following five years. The program was bound to be highly controversial, but promised to unlock a great portion of the nation's energy resources.

Development of offshore exploration, drilling, and production technology rivals that of space exploration. As a result, the 1980s are the "deep water decade." Operations already can be carried out in 6,600 feet of water, and the industry no longer sees 13,000 feet of water as an obstacle to exploratory drilling.

Development of drillships that do not need anchors was to offshore oil and gas exploration what landing a man on the moon was to space exploration. With oil and gas being sought for in deeper and deeper waters where conventional drillships could not be anchored to the ocean floor, or the water was too deep for drilling platforms, technology was developed to use a system of computerized underwater thrusters to keep drillships in place. Once the dynamically positioned drillship is on location, ocean floor equipment is set and drilling starts; information about the exact location of the drilling hole is beamed directly from the ocean floor to computers on board, through sensors in the hull of the ship. From that point on, it is virtually locked into position. Every time wind, waves, or ocean currents push the ship off target, the computer-activated thrusters automatically push it back on track. This dynamic positioning system is so accurate that the drillship can hover over the same spot almost indefinitely.

The search for oil and gas offshore Atlantic is another classic story of frustrations in opening up new oil provinces. In 1978, when Texaco announced a "significant gas discovery," there was great excitement. However, by the end of 1982, twenty-seven wells had been drilled in the area by nine companies, but found no commercial production. Drilling in Georges Bank, offshore Massachusetts, was equally disappointing. By 1982, eight dry holes were drilled by five companies, bringing the cost of fruitless offshore Atlantic drilling in the two areas to $2.9 billion for thirty-five exploratory wells and no commercial production. Oil companies have not yet written off the area's potential, but will have to invest more huge amounts of capital to continue exploration.

Development of ice technology is the newest challenge to offshore explorers in Alaska. The U.S. Geological Survey estimates the oil potential of all Alaskan offshore areas to be between 7 and 32 billion barrels and offshore gas potential to be between 30 and 97 trillion cubic feet. The Beaufort Sea in the Arctic Ocean off the North Slope, where the giant Prudhoe Bay field was found, is considered the most promising U.S. offshore exploration frontier. Exploratory drilling has made significant discoveries in shallow waters close to shore, and the industry

TEXACO INC.

Discoverer Seven Seas, *a dynamically positioned drillship, owned by Sonat Offshore Drilling, is shown heading for location offshore Newfoundland, Canada, where it set a world record in 1980 by drilling in almost 5,000 feet of water. The $35 million Texaco wildcat, drilled to 20,023 feet below the ocean floor, was a dry hole.*

plans an intensive search for giant fields during the 1980s.

The government permits drilling only between November 1 and March 31 when the sea is frozen solid on the theory that oil spills from drilling accidents pose less danger when the sea is frozen. Exploration wells, costing $20 million apiece, are drilled on man-made gravel islands. Holes are cut in the ice, and gravel, hauled over ice roads, is dumped into them. A typical island in sixty-three feet of water required 5 million cubic yards of gravel, measured a half-mile in diameter and took two winter seasons to build. A production island, which must accommodate several wells for the life of the oil field, along with pipelines and other equipment, might require six times as much gravel. Ice engineers are developing technology for exploration and production platforms needed in Arctic waters too deep for artificial islands. Seismic operations over millions of acres over the sea are conducted by vibroseis trucks on ice during the winter and by boat during the brief summer.

Cognac has set many world records in addition to being the world's tallest offshore platform. The first three-part platform weighs 59,000 tons—the world's heaviest. It is the first complete offshore platform construction which used underwater hammers. New endurance records were set by divers working in 850 to 1,025 feet of water. When first installed in 1978, it was used for three years to drill sixty-two wells, each to 14,000 feet depth. Then, converted to a production platform, it began production in 1982 of 45,000 barrels of oil a day and 87 million cubic feet daily of natural gas. *Cognac* cost $275 million compared with the first platform in the Gulf of Mexico in 1947 which cost $230,000.

Amoco's Houston exploration staff, below, intently follow bidding on federal offshore tracts in the Texas Gulf Coast as they listen to the broadcast lease sale in New Orleans.

STANDARD OIL (INDIANA)

MOBIL

The vibroseis truck above is recording seismograph data in the Beaufort Sea in the Arctic Ocean. Signals are shot into the seabed through four feet of ice on top of up to fifty feet of seawater.

EXXON

The exploratory drilling rig above is on an artificial gravel island in the frozen Beaufort Sea.

CONTINENTAL SHELF ASSOCIATES, INC.

In 1977 Mobil found a 90-billion-cubic-feet gas field in the Gulf, 110 miles southeast of Galveston, Texas, but due to environmentalists' actions was not issued permits for development drilling until 1981. The field was near the Flower Garden Banks, the northernmost living coral reefs in the Gulf. According to lease terms, Mobil spent half a million dollars on monitoring the ecology of the reef, as shown at left, before and after exploratory drilling. The studies showed drilling did not affect the reefs, but it took four years of hearings demanded by environmental groups before development could begin.

The world's tallest offshore platform is Shell Oil Company's Cognac, offshore Louisiana in the Gulf of Mexico. At left, with two drilling rigs in 1978, it stood 1,265 feet high in 1,025 feet of water. At right, as a production platform in 1982, with flure tower atop, it stands 1,406 feet high.

TEXACO INC.

To Ocean Victory, the drilling rig above went the victory of discovering the first offshore Atlantic hydrocarbons in a $9.2 Texaco wildcat in the Baltimore Canyon in 1978, starting the difficult search to prove the oil and gas potential of this frontier area.

One of the latest advances in underwater technology is the one-atmosphere diving suit, shown above.

Moving into deeper waters involved the invention of subsea massive blowout preventers, shown above.

LOOP, America's first superport, owned by Marathon, Texaco, Shell, Ashland, and Murphy, is a unique $700 million system providing a safer, more economical way to handle one-fourth of the nation's oil imports at the rate of 1.4 million barrels daily. No other offshore oil structure approaches the scale and sophistication of LOOP. Its nerve center, a computerized marine terminal eighteen miles offshore with a single-point mooring buoy for supertankers, pumps oil through forty-seven miles of pipelines offshore and onshore to storage caverns in a salt dome, which then feeds the oil into national pipeline networks that handle one-fourth of the nation's refining capacity, some of them as far north as the Great Lakes. The total system ranks as a spectacular engineering feat that experts say exceeds in scope and complexity the building of the Alaska pipeline. LOOP is a monument to unusual cooperation between the oil industry and environmentalists. LOOP officials consulted environmentalists in the design phase and allayed their fears of oil spills by putting storage underground instead of above, and helped preserve the Louisiana wetlands by filling in the ditches dug during pipeline construction. Until LOOP was constructed no American port could receive supertankers. It also reduces the danger of oil spills from offloading supertankers offshore into smaller tankers.

The Texaco Caribbean, *above, unloads Saudi Arabian crude in April 1981, at the inauguration of Louisiana Offshore Oil Port, or LOOP, America's first superport for directly unloading oil from supertankers.*

LOOP's superport offshore platform complex, below, pumps oil from supertankers through submarine pipelines to a vast pipeline network on shore.

Technology for laying underwater pipelines from offshore production to onshore storage and pipelines has progressed as rapidly as that for offshore drilling and production. One of the most sophisticated semisubmersible pipelay barges in the world is Brown & Root's Semac I, shown at left, which can lay pipelines in waves up to fifteen feet.

BROWN & ROOT, INC.

One of the biggest potential new giant oil fields in the lower forty-eight states is offshore northern California near the place where the Santa Barbara channel opens up into the Pacific. Standard Oil of California and Phillips Petroleum, in a joint venture, discovered it in late 1981. In order to carry out development drilling they contracted the huge, new semisubmersible drilling rig, Zapata Concord, shown below, being towed in 1982 through the Strait of Magellan at the southern tip of South America on the world's longest drilling rig tow—14,060 miles from Mobile, Alabama, to Santa Barbara. It began drilling on the offshore tract which drew the highest bid in U.S. lease sales—a record 333.6 million dollars bonus in mid 1981. The wildcat discovered a supergiant oil field which may be comparable to Alaska's Prudhoe Bay Field.

PHILLIPS PETROLEUM

MOBIL   MOBIL

Mobil Search, *the largest and most advanced geophysical vessel in the world in 1982, shown above left, has a computerized data acquisition system to handle 450 channels of seismic data, process it on board, and transmit it to headquarters in Dallas. At right are the reels of hydrophone cables which are towed steadily behind the vessel. As the ship moves along, air guns are fired automatically to send seismic waves thousands of feet below the ocean floor and reflect the contour of the rock formations through the hydrophones.*

The largest international conferences in the world are those displaying the latest advances in offshore oil and gas exploration and drilling. Below is the 1981 conference in Houston, Texas.

MARATHON

Since the discovery of the first commercial oil well in Pennsylvania in 1859, the energy history of the world has been known as "The Petroleum Age." But beginning in the 1970s it became increasingly evident that the world was entering "The Natural Gas Age." New deep gas discoveries, new research and technology, provide proof that natural gas is becoming a major, and perhaps the largest, future source of world energy.

Natural gas price controls in the United States established in 1954 provided little incentive to drill for natural gas in its own right. However, since 1969, independent pioneers have demonstrated that there is an astounding amount of natural gas to be found at greater depths than conventional oil resources. The "father of deep natural gas" is Robert A. Hefner III, an Oklahoma City geologist, who, in the tradition of the great wildcatters and scientists, pioneered a new frontier for the discovery of huge deep gas resources.

In the late 1950s, as a geology student at the University of Oklahoma, Hefner studied the stratigraphy of the Anadarko Basin, the vast L-shaped structure covering thousands of square miles, extending from the Texas Panhandle across the Oklahoma Panhandle and into Kansas on whose shallow rims were some of the world's largest proven gas reserves. It was known that the Anadarko Basin was one of the deepest in the United States with in excess of 40,000 feet of sediments, or 24,000 cubic miles of untested potential for hydrocarbons. The deeper wells are drilled, the greater is the likelihood of finding natural gas. However, accepted geological thinking held that at depths from 15,000 to 50,000 feet the weight of the overburden would have crushed any potential for big natural gas reservoirs. Robert Hefner did not believe this. With two partners, he formed the GHK Companies and managed to find financing for his theory. His first deep test was a dry hole, but in 1969, after two years of overcoming immense drilling problems and making many technological breakthroughs, he drilled a well to 24,454 feet, then the second deepest well ever drilled in the world, and found a huge gas reservoir.

Robert Hefner's pioneering encouraged new deep drilling in search of gas to be sold in the non-price-controlled intrastate markets. It was not until

GHK COMPANIES

Robert A. Hefner III, above, the Oklahoma geologist and wildcatter who is the father of deep natural gas, has pioneered the transition of the Petroleum Age to the Natural Gas Age.

the passage of the 1978 Natural Gas Policy Act, phasing out price controls by 1985, and eliminating price controls on production below 15,000 feet, that the deep gas boom began which marked the beginning of "The Natural Gas Age."

Natural gas drilling began adding more energy equivalent to the nation's proven energy reserves than oil drilling. The potential of the Anadarko Basin alone is estimated to be the energy equivalent of the discovery of six giant oil fields. In the late 1970s and early 1980s huge discoveries of deep natural gas also were made in the Gulf Coast; the Williston, Appalachian, and Michigan Basins; the Rocky Mountain Overthrust Belt; and the Deep Tuscaloosa Trend in

PHILLIPS PETROLEUM

The new search for deep natural gas led to the discovery in 1975 by Socal of a major natural gas province, called the Tuscaloosa Trend, stretching 200 miles across Louisiana from Mississippi to Texas. Deep natural gas is being developed from 16,000 to 22,000 feet, as shown above in Louisiana marshland. The trend is estimated to contain 2.5 trillion cubic feet of natural gas and 100 million barrels of condensate—a form of natural gas.

A wide variety of massive drilling bits are needed to drill a 24,482-foot natural gas well at Mills Ranch Field, Texas, as shown below.

The newest use of natural gas is to replace gasoline in vehicles, as shown below. In 1982, over 20,000 vehicles in the United States had been converted to use either compressed natural gas or gasoline. It is particularly economic for operating fleet vehicles.

SOCAL

DUAL FUEL SYSTEMS INC.

Louisiana. The spectacular new results in the United States of an abundance of gas at depth are being repeated worldwide—in the Soviet Union, Iran, Indonesia, Qatar, Argentina, West Germany, Taiwan, and other countries.

Conventional geological theories have assumed that natural gas and other hydrocarbon deposits were produced solely by the decomposition of plant and animal remains from previous eras and are organic. However, Dr. Thomas Gold, Director of the Center for Radiophysics and Space Research at Cornell University, a leading cosmologist concerned with the formation and development of planets, is the originator of the new theory that methane, or natural gas, is a major nonbiologic component of the formation of the earth and that vast natural gas resources exist at depth all over the world.

There is an abundance of shallow gas yet to be found in the United States and also huge supplies to be found in tight formations and potential traps concerned with fractured rocks, impact meteorite craters, and other types of reservoirs.

Natural gas promises to be the newest, most exciting, and productive energy frontier.

While American oil and gas explorers are pioneering frontier areas for undiscovered oil and deep natural gas onshore and offshore, American technologists are pioneering another huge energy frontier—the art of recovering the estimated 235 billion barrels of already discovered oil known to be in place in the United States, but unproducible before the development of an expensive new technology called enhanced recovery. Average recovery of U.S. oil reserves is only about 33 percent. Water flooding and gas reinjection greatly increase recoveries, but new methods of steam stimulation and the use of chemicals have enormous potential in recovering known, but heretofore unrecoverable, oil, both onshore and offshore.

GHK COMPANIES

The world's largest land rig, at right, towers over a cow pasture in Oklahoma's Anadarko Basin, proving ground for vast, new, deep natural gas resources. Operated by Robert A Hefner III's GHK Companies, this Parker Drilling Company rig is drilling the No. 1-1 Robinson well in search of gas at a depth of 33,000 feet. Taking two years to drill at a cost of $20 million, when completed in 1983, it will be the world's deepest well. A 525-square-foot American flag waves over the derrick. The rig is capable of drilling to 50,000 feet—the yet unprobed depths of the promise of tomorrow's natural gas resources.

STANDARD OIL (INDIANA)

A woman holds a container of heavy crude oil upside down, at left, demonstrating how inert and incapable of flow crude oil is in underground rocks once the natural gas mixed with it has escaped or been produced.

Steam flooding is currently the primary enhanced recovery method. Water converted into steam is injected through wells into oil formations. Steam causes the oil to thin, like melting wax, and flow to producing wells to be pumped. Below left, is Texaco's San Ardo, California, steam flood, one of the world's largest. Below right, is Socal's steam flooding through huge pipes in central California.

TEXACO INC.

SOCAL

TEXACO INC.

Chemical solvents are being used to wash out inert oil in the ground like detergents are used to wash dirt out of laundry. A major technique is pumping carbon dioxide into injection wells to produce more oil, as shown above in Texaco's Louisiana pilot project.

The scanning electron microscope, shown below, is a powerful tool, magnifying objects 100,000 times, to enable geologists to study rock formations to determine which method of enhanced recovery is the best to use.

TEXACO INC.

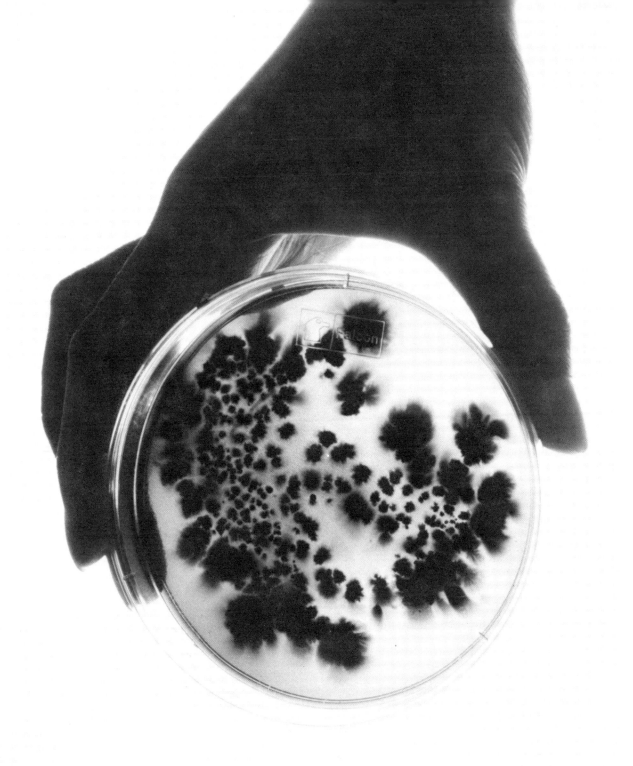

SOCAL

A glimpse of the future lies in the microscopic view above of microorganisms in a petri dish in research being carried out by Cetus Corporation of California on developing living materials to be used in enhanced oil recovery instead of chemicals. The emerging field of industrial microbiology is another demonstration of the universal truth that nature's resources are unlimited as long as the minds of men are constantly probing to find new ways to use and develop them.

ENERGY FUTURE

America's oil and gas history since 1859, has demonstrated some principles of enduring importance which can guide the nation to energy self-sufficiency and an abundance of affordable energy.

As the great wildcatters proved, and continue to prove, a wildcatter can't quit, and exploration must be a continous process. Despite informed estimates of how much oil and gas remains to be found, explorationists and technologists constantly expand the nation's, and the world's, potential by discoveries and new techniques.

Man's political activities, not nature, provide the greatest obstacles and limits to energy development and use. America's series of energy crises proved that price controls and excessive regulations induce scarcity of supplies. Abundance is created only in a free marketplace by a competitive, free enterprise society relying on the incentive of profits. Of equal importance is access to potential resources onshore and offshore. The industry has proved that modern exploration and development of oil and gas resources pose no great threats to the environment. The nation's long-range economic development depends upon wise, multiple use of its public lands. The nation cannot afford stifling its economic growth by actions of overzealous environmental groups who are more concerned about the potential, and generally unwarranted, damage to the physical environment than they are about the welfare of mankind.

Although conventional oil and gas resources are finite, there is a consensus of scientific opinion that they will continue to provide a major share of energy supply for at least the next 100 years. During this period, the energy mix will change. Synthetic fuels from coal and oil shales; nuclear, solar, and wind power, will begin providing more energy as it becomes economically and technologically feasible for them to play a greater role. As an abundance of new energy supplies are developed, low-cost energy will once more help pave the way to world progress.

Energy is the key to creating wealth, not just for the few, but the many. Today's energy problems and fears seem unnecessarily exaggerated in view of the abundance of energy nature has placed here for use. The atom's potential is such that a handful of snow, if entirely converted to energy, could keep a home furnace running continuously for some 25,000 years.

The last half of the twentieth century is the period of the world's greatest knowledge explosion. Scientists say that by the mid-eighteenth century, mankind's total fund of knowledge doubled from what it was at the birth of Christ. It doubled again by 1900 and again by 1950. *But*, the next two *doublings* of mankind's knowledge took place in 1960 and 1968. The rapid advance of technology is providing another doubling.

However, knowledge explosions are not enough. Human ingenuity, worldwide, faces its greatest challenge—developing the political skills to put knowledge to use wisely to end world scarcities and create abundance for all.

Selected Bibliography and Recommended Readings

Harold F. Williamson and Arnold R. Daum, *The American Petroleum Industry, 1859-1899, The Age of Illumination*, Northwestern University Press, 1959; Allan Nevins, *Study In Power* (2 vols., biography of John D. Rockefeller), Scribner, 1953; Ralph W. Hidy and Muriel E. Hidy, *Pioneering in Big Business* (vol. 1, history of Standard Oil Company of New Jersey, 1882-1911), Harper, 1955; George Sweet Gibb and Evelyn H. Knowlton, *The Resurgent Years* (vol. 2, history of Standard Oil Company of New Jersey, 1911-1927), Harper, 1956; Henrietta M. Larson and Kenneth Wiggins Porter, *History of Humble Oil & Refining Company*, Harper, 1959; Gerald T. White, *Formative Years in the Far West*, (history of Standard Oil Company of California and predecessors through 1919), Appleton, 1962; Kendall Beaton, *Enterprise in Oil*, (history of Shell in the United States), Appleton, 1957; Paul H. Giddens, *Standard Oil Company (Indiana)*, Appleton, 1955; Earl M. Welty and Frank J. Taylor, *The Black Bonanza* (history of Union Oil Company of California), McGraw-Hill, 1958; Sam Mallison, *The Great Wildcatter* (biography of Mike Benedum), Education Foundation of West Virginia, 1953; John Joseph Mathews, *Life and Death of an Oilman* (biography of E. W. Marland), University of Oklahoma Press, 1952; James A. Clark and Michel T. Halbouty, *Spindletop*, Random House, 1952; Wallace E. Pratt, *Oil in the Earth*, University of Kansas, 1943; Ruth Sheldon Knowles, *The Greatest Gamblers: The Epic of American Oil Exploration*, University of Oklahoma Press, 1978; Ruth Sheldon Knowles, *America's Energy Famine: Its Cause and Cure*, University of Oklahoma Press, 1980.

Index

Advertising: kerosene lamp, 8; Standard Oil, 11; Texaco, 71
Aldrin, Edwin E.: 108
Alyeska Pipeline: 111, 113, 120-27
American Association of Petroleum Geologists: 53-54, 146
American Petroleum Institute: 114
Amoco Oil Co.: 101, 137
Anglo-Iranian Oil Co.: 86-87
Apollo 11: 108
Arab-Israeli War, 1973: 118
Arab Oil Embargo: 118-19, 127
Arco (Atlantic Richfield): 111, 120, 126
Armstrong, Neil: 108
Army-Navy Petroleum Board: 75
Ashland Oil Co.: 153
Automobiles: first American, 15; 31-34, 45, 49, 58, 71, 89, 117; see Henry Ford
Aviation: 35, 58, 61, 80-81

Bahrein: 86-87
Benedum, Mike: 56-57
Bigelow, Dr. Henry: 12
Bissell, George: 3, 5
Burton, Dr. William M.: 49, 72
Bush, George: 96
Butadiene: 79, 85

Canfield, Charles: 16
Carter, James Earl, Jr. (Jimmy): 128-29
Cetus Corporation: 162
Cities Service: 41-43, 58-60, 94
Coal: 89, 91, 133-34, 145
Cognac: 149, 151
Colombia: 136
Conoco: 53, 68, 93, 95, 103, 105, 137, 141, 144
Conservation: 68-69, 73, 119, 128-31
Continental Shelf and Slope: 145-47
Cram, Ira. H.: 115-16
Cullinan, Joseph S.: 23
Curzon, Lord: 45

DeGolyer, Dr. Everette Lee: 22, 65, 86-87
Doheny, Edward L.: 15-16, 55

Drake, Col. Edwin L.: 3, 5, 7
Drilling: dynamically positioned drillship, 147-48; first cable tool rigs, 5-7; first drill ship, 95; first offshore jackup rig, 96; first rotary rigs, directional, 67-68; first submersible drilling barge, 70; in Los Angeles, 104; rotary, for oil, 19, 21, 28
Dubai: 137
DuPont: 97-98
Duryea, Charles: 15

Eastman, John: 67
Ecuador: 136
Egypt: 136
Enhanced Recovery: 159-61
Environment: oil industry protection measures, 93, 104-05, 112, 114, 120, 124-25, 145, 150
Environmentalists: 112-13, 115, 120, 127, 143, 146, 150, 153
Exxon: 90, 92, 93, 111, 115, 116; *also see* Humble Oil Co.

Fall, Albert: 55
Federal land ownership: 145
Federal Power Commission: 91
Fires: 25, 38, 67
Ford, Gerald R.: 128
Ford, Henry: 17, 32

Galey, John H.: 20, 23
Gasoline: 12-13, 33, 45, 49, 58, 61, 72, 80-81, 83, 85, 117, 119, 128-29
Geology: 41-43, 53-55, 57-60, 63, 65-67, 71
Getty, J. Paul: 37
Gold, Dr. Thomas: 159
Gould, Charles N.: 41, 57
Guffey, Col. James M.: 20, 23
Gulf Oil Corporation: 20, 23, 27, 31, 77, 87, 111, 136

Hamill, Al and Curt: 20-22
Hefner, Robert A., III: 157
Higgins, Patillo: 19, 21
Houdry, Eugene: 72
Hughes Tool Co.: 28
Humble Oil Co.: 53-54, 63, 67, 70, 104

167

Ickes, Harold L.: 75-76
Imports: 51, 91-92, 111, 115-17, 128-31
Independents: 91, 93, 111, 120, 129, 141
Indonesia: 136
Interstate Oil Compact: 68
Iraq Petroleum Co.: 51

Joiner, "Dad" Marion Columbus: 63-64

Kier, Samuel: 4, 5
Kuwait: 87

Landsat: 109
Laniscot I: 94
Libya: 115-18
Lindbergh, Charles A.: 58
Liquified Natural Gas (LNG): 127
Liquified Petroleum Gas (LPG): 72, 88-89
Lloyd, Dr. A. D.: 63-64
Louisiana Offshore Oil Port (LOOP): 153-54
"Louisiana Shrimp & Petroleum Festival," Morgan City: 105-06
Lucas, Capt. Anthony F.: 19-22, 23

Magnetometer: 77
Marathon Oil Co.: 144, 153
Marland E. W.: 50, 53, 68-69
Marland Oil Co.: 50, 51, 68
Mellon, Andrew W.: 20, 23
Mobil Oil Corp.: 72, 104, 111, 150, 156
Murphy Oil Co.: 153

National Audubon Society: 104-05
National Environmental Protection Act of 1970: 113, 115-16, 145-46
National Petroleum Council: 115-16
National Petroleum War Service Committee: 46
Natural gas: 39, 53, 60, 70, 73, 85, 89, 91, 128-30, 140 157-59
Natural Gas Pipeline Company of America: 70
Natural Gas Policy Act: 157
Nimitz, Fleet Admiral Chester W.: 75, 80
Nixon, Richard M.: 116-17, 128
Norway: 133
Nuclear energy: 133

Offshore: exploration, 91-96; 104, 111, 115-16, 135, 145-56, 159; lease sales, 91, 93, 145-46, 149, 155; oil spills, 111-14; technology conference, 156
Ohio Oil Co.: 57

Oil companies: *see* Amoco, Arco, Anglo-Iranian Oil Co., Ashland Oil Co., Cities Service Oil Co., Conoco, Exxon, Gulf Oil Corp., Humble Oil Co., Iraq Petroleum Co., Marathon Oil Co., Marland Oil Co., Mobil Oil Corp., Murphy Oil Co., Ohio Oil Co., Phillips Petroleum Co., Shell Oil Co., Standard Oil, Standard Oil of California (SOCAL), Standard Oil of Indiana, Standard Oil of Ohio (SOHIO), Sun Oil Co., Texaco, Union Oil Co.
Oil fields: Alaska: Kenai Peninsula, Cook Inlet, 127; Prudhoe Bay, 110-11, 120, 123, 125-27
  Amazon Valley: 136
  California: Hondo, 115; Huntington Beach, 54-55; Lakeview, 36; Los Angeles Basin, 54-55, 103; Los Angeles City, 15-16; Signal Hill, 55
  Canada: Le Duc, 92
  Dubai: 137
  Indonesia: Minas, 136
  Kansas: Augusta, 41; El Dorado, 41-42
  Kuwait: 87
  North Sea: Norway, 133, 135
  Oklahoma: Burbank, 50, 53; Cushing, 37-41; Edwards, 65; Glenn Pool, 31; Oklahoma City, 59-60, 68; Seminole, 59
  Pennsylvania: 3-7
  Saudi Arabia: 87
  Texas: Big Lake, 56-57; Borger, 57; Conroe, 67-68; East Texas, 62-65; King Ranch, 70; Mexia, 52-54; Ranger, 45, 47; Sour Lake, 23-27; Spindletop, 18-27; Yates, 57
  Venezuela: Lake Maracaibo, 76
  Wyoming: Salt Creek, 55
Oil storage: 9, 25, 137, 153
Oil tank cars: 9
Oil wells: Burbank, Okla., 50, 53; California's first commercial, 13-14; Drake Well, Pa., 3, 5, 7, 14, 146; East Texas, Tex., 62-65; Lakeview, Cal., 36; LeDuc, Canada, 92; Mexia, Tex., 52-54; Minas, Indonesia, 136; Prudhoe Bay, Alaska, 110-11, 120, 123, 125; Santa Rita, Big Lake, Tex., 56-57; Spindletop, Tex., 18-27; world's deepest, Anadarko Basin, Okla., 157, 159
Organization of Petroleum Exporting Countries (OPEC): 117-18, 128-29
Osage Indian Tribe: 50-53
Overthrust Belt: 141-46, 157

Petrochemicals: 97-103, 108
Petroleum Administration for War: 75-76, 85, 86
Phillips, Frank: 52, 53

Phillips Petroleum Co.: 52, 53, 58, 61, 80, 99, 102, 127, 132-35, 155

Pipelines: Alyeska Pipeline, Alaska, 111, 113, 115, 120-27; Big Inch & Little Big Inch, 77-78, 91; first cross country, natural gas, 70; first pipeline company, 9; military mobile, 78; natural gas, 89, 91; Tapline, Saudia Arabia, 115; Texas to Oklahoma, 31; Trans-Andean and Trans-Ecuadorian, 136; underwater, 154-55; West Texas, 90

Potential Gas Committee: 139

Pratt, Wallace E.: 53-54, 63, 67, 70

Price controls: 91, 111, 117, 128-30, 157

Products: asphalt, 28, 33, 75, 86, 132; cold cream, 34; fuel oil, 13, 26, 71, 89; gasoline, *see* Gasoline; harness oil, 12; kerosene, 2-3, 7-8, 11-12, 97; medicinal, 5, 12; naphtha, 12-13; paraffin wax, 13, 86; petrochemicals, *see* Petrochemicals; petroleum jelly, 12-13; soap, 34; synthetic rubber, 75, 81-83; toluene, 75, 79

Profits, oil company: 118

Pulp weeklies: 29-30

Reagan, Ronald: 129

Refineries: Exxon, 90; Gulf Refining Co., Port Arthur, Tex., 23, 31; Newhall, Cal., 14; Standard Oil of Indiana, 48-49, 105; Standard Oil of Ohio, 7; Sun Oil Co., 72; Texaco, Port Arthur, Tex., 26

Rockefeller, John D.: 7-8, 31, 37

Roosevelt, Franklin D.: 76

Roosevelt, Theodore: 36-37

Salt domes (plugs): 19, 26, 43, 65, 67

Santa Barbara oil spill: 111-14

Satellite: 109

Saudi Arabia: 86-87

Schlumberger electric well-logging: 67

Seismograph: 65-67, 93-94, 139-42, 149, 150, 156

Service stations: 33-34, 58, 71, 83

Shale oil: 133, 145

Shell Oil Co.: 55, 58, 67, 78, 95, 104, 107, 149-50, 151, 153

Sherman Anti-Trust Act of 1890: 8, 37

Silliman, Professor Benjamin: 2-3, 14

Sinclair, Harry F.: 31-32, 37, 46, 55-56

Smith, Levi: 56-57

Smith, Sidney S.: 78

Smith, Uncle Billy: 5, 7

Solar energy: 133

Song, American Petroleum: 10

Standard Oil: 8, 11, 19, 23, 31, 37, 38, 53

Standard Oil of California (SOCAL): 54, 86, 103, 136, 155, 158, 160

Standard Oil of Indiana: 16, 48, 49, 101, 105, 139

Standard Oil of Ohio (SOHIO): 8, 97, 99, 126

Stewart, Lyman: 13-16, 36

Strake, George: 67

Sun Oil Co.: 23, 72

Synthetic fibers: 97-99, 108

Synthetic fuels: 133

Synthetic rubber: 75, 79, 84, 85, 97

Tankers: 27, 36, 38, 44, 74, 76-77, 80, 137, 153-54

Teapot Dome: 55

Texaco: 23, 28, 31, 44, 70, 76, 86, 104, 111, 132, 134, 136, 140, 147-48, 152-54, 160-61

Toluene: 75, 79

Tractors: 45, 71, 87-88

Tuscaloosa Trend: 157-59

Union Carbide: 99

Union Oil Co.: 15, 36, 95, 100, 104, 107

United States Geological Survey: 139, 145, 147

United States Technical Oil Mission: 86

Venezuela: 76

Vibroseis: 138, 141, 149-50

Williams Brothers Pipeline Co.: 136

"Windfall Profits" tax: 129

Woods, Charles Lewis ("Dry Hole" Charlie): 36

World War I: 38, 44-49

World War II: 72, 74-88

Wright, Orville and Wilbur: 17, 35

About the Author

The granddaughter and daughter of independent oilmen, Ruth Sheldon Knowles is a seasoned oilwoman in her own right. She is an internationally known petroleum specialist, leading historian of the oil and gas industry, writer, foreign correspondent and lecturer. She has made eleven trips around the world studying oil exploration and production in the world's major oil centers.

Mrs. Knowles has served as a petroleum consultant to the United States, Mexican, Venezuelan and Indonesian governments. In 1941, she was appointed by Secretary of Interior Harold L. Ickes to be a petroleum specialist on his staff. He sent her to South America to make the first United States government survey of oil fields and refineries. She was a special consultant to the Venezuelan government on its petroleum law. She spent much time in Cuba on oil exploration projects.

In addition to her books, Mrs. Knowles is a frequent contributor to leading popular magazines, newspapers and professional journals. She has written radio programs for "Voice of America," and a series of radio programs concerning the history of Middle East oil that has been broadcast in Arabic in Middle East countries. She wrote, produced, and narrated a documentary film on Pertamina, Indonesia's national oil company.

The American Women of Radio and Television selected Mrs. Knowles as "Woman of the Year" in Oklahoma in 1961, and in 1962 she was selected as Oklahoma's outstanding woman journalist by Theta Sigma Phi, the national professional fraternity for women in journalism. She is a founder and director of the Women's Economic Round Table.

Mrs. Knowles, who resides in New York City, is the mother of four children.